Sustainable Corrosion Inhibition Using Agricultural Waste

This book discusses corrosion and inhibition using agricultural waste including the impact of corrosion on key emerging technologies such as 3D printing, clean energy, smart coating, and machine learning via environment, sustainability, and governance and economies (ESG) approach. The advantages and disadvantages of using this eco-friendly, sustainable natural product as a corrosion inhibitor over other commercially available corrosion inhibitors is discussed.

Features:

- Discusses the concept of Industry 4.0 in corrosion inhibition technology.
- Explains how agricultural wastes are used in solving global corrosion challenges that aim to demystify machine learning, artificial intelligence, and waste to wealth, in different industries.
- Reviews in-depth inhibitor application in solving global challenges of housing, transport, oil and gas, among others.
- Explores impact of corrosion on the environment, sustainability, and governance and economies.
- Examines corrosion and 3D printing focusing on history, materials, manufacturers, and trends.

The book is aimed at researchers and graduate students in corrosion, materials science, and waste processing.

Sustainable Corrosion Inhibition Using Agricultural Waste

An ESG Strategy

Omotayo Sanni, Kingsley Ukoba, Jianwei Ren, and Tien-Chien Jen

CRC Press is an imprint of the
Taylor & Francis Group, an informa business

First edition published 2025
by CRC Press
2385 NW Executive Center Drive, Suite 320, Boca Raton FL 33431

and by CRC Press
4 Park Square, Milton Park, Abingdon, Oxon, OX14 4RN

CRC Press is an imprint of Taylor & Francis Group, LLC

ISBN: 9781032533230 (hbk)
ISBN: 9781032578187 (pbk)
ISBN: 9781003441151 (ebk)

DOI: 10.1201/9781003441151

Typeset in Times
by Newgen Publishing UK

Contents

Preface

Corrosion can be slowed down in different ways, including corrosion inhibitors, cathodic protection, metal impurity reduction, application of surface treatment methods, and the incorporation of appropriate alloys. Among these methods, using environmentally sustainable corrosion inhibitors has been validated to be one of the cheapest and simplest method of corrosion protection and prevention in different environments. Corrosion inhibitors slow the metallic dissolution rate, minimizing economic losses caused by the metallic losses on industrial tools or surfaces. Traditional inorganic and organic inhibitors are expensive, and therefore, recently, research and development on green corrosion inhibition have been gaining particular attention because of increasing ecological awareness and strict environmental regulations. Several environmentally friendly corrosion monitoring practices and inhibitors have been developed and implemented in the last few years.

Many agricultural wastes have not been timely and effectively recycled, resulting in severe problems of environmental pollution and waste of resources. To help support the effort, various governmental and non-governmental agencies established funding for comprehensive agricultural waste utilization technology. The extensive utilization of agricultural waste has attracted worldwide attention. Using agricultural waste is an effective way to improve resource utilization efficiency and products' added value. Therefore, this book assembles the use of agricultural wastes as corrosion inhibitors and their latest applications. Throughout this book, agricultural wastes as environmentally sustainable corrosion inhibitors are cost-effective and can control and retard corrosion in different aggressive environments. Environmental regulations in industrialized countries are raising the pressure to remove many compounds used broadly in the industry to protect against corrosion. Environmentally sustainable corrosion inhibitors are currently emerging, oriented toward minimizing the environmental impact and providing effective corrosion inhibition. Environmentally sustainable corrosion inhibitors contain agricultural wastes, natural products, plant extracts, and synthetic nontoxic materials that can replace the toxic marketable products still used in many industries worldwide. This book offers the synthesis, characterization, inhibition mechanism, and applications of environmentally sustainable corrosion inhibitors in industry. The examples provided in this book include areas such as agricultural wastes, the environment, oil and gas, etc. This book is divided into two parts, each containing different chapters. Part 1 "Corrosion Inhibitors: Fundamental Concepts" covers topics such as green chemistry principles and green corrosion inhibition, green corrosion inhibition practices, classification of corrosion inhibitors, toxicity of corrosion inhibitors, introduction to sustainable corrosion inhibitors, industrial applications of corrosion inhibitors, describes the journey from traditional to agricultural waste as corrosion inhibitors, economic and industrial opportunities, agricultural wastes extract as sustainable corrosion inhibitors, applications of agricultural wastes as organic green corrosion inhibitors, modern testing and analyzing techniques in corrosion inhibition, analysis in corrosion, commercialization of environmentally sustainable corrosion inhibitors. Part 2 "Corrosion and Emerging Technologies" describes the introduction to emerging technology and 4IR, effect of corrosion on emerging technology, key emerging technologies, clean energy, corrosion, and additive manufacturing, background of three-dimensional (3D) printing, concept of 3D printing, corrosion and impact on 3D printing, corrosion impact on Biomedical 3D printing application, corrosion impact on automotive, aviation and aerospace in 3D printing application, corrosion impact on 3D printing application in space, corrosion impact on construction in 3D printing application, corrosion impact on warfare in 3D printing application, corrosion impact on food in 3D printing application, and corrosion, industrialization, climate change and energy security in the Global South, which entirely focuses on the Global South, conceptual and theoretical perspectives of industrialization, climate change, and energy security, perspectives and trends of industrialization, climate change, and energy security in the Global South issues and challenges of industrialization, climate change, and energy security in Global South, the future outlook of industrialization, climate change, and energy security in the Global South. Corrosion and smart coating that details water sanitation and hygiene (WASH), state of WASH and smart coatings.

Overall, this book is written for researchers in academia and industry, corrosion engineers, scientists, and students of materials science, applied and engineering chemistry. The editors and contributors are well-known researchers, scientists, and true professionals from academia and industry.

About the Authors

Omotayo Sanni is a researcher in the Department of Chemical Engineering, University of Pretoria, Pretoria, South Africa. She obtained a doctoral degree in Chemical Engineering from the Tshwane University of Technology, Pretoria, South Africa. She has demonstrated expertise leading to the publication of several reputable, peer-reviewed journal articles and presentations at national and international conferences.

Kingsley Ukoba is a lecturer and researcher in the Mechanical Engineering Science department of the University of Johannesburg, South Africa. He obtained a doctoral degree in Mechanical Engineering from the University of KwaZulu-Natal, Durban in South Africa, graduating among the top 15 researchers. He coordinates the smart energy group for JENANO headed by Professor Jen. He is among the authors for the African Integrated Assessment report by the United Nations Environment Programme, African Union Commission, CCAC and Stockholm Environment Institute (SEI) joint publication. He is also a 2022 Engineering for Change (E4C) fellow sponsored by American Society of Mechanical Engineers (ASME) to support an Impact Project around Climate Action. He has authored books, book chapters and journal articles, presented in conferences and serves as a reviewer for high impact journals and conferences.

Jianwei Ren is currently working as a full professor at Department of Chemical Engineering, University of Pretoria (UP). His research interests cover automatic data acquisition, process control, materials science and advanced manufacturing, H_2 and fuel cell technologies, ALD technologies, system integration, waste and drinking water treatment for sustainable environment. He has published 160 journal articles, 50 conference proceedings, two patents, 13 book chapters and presented in over 60 international and local conferences.

Tien-Chien Jen is the head of department of Mechanical Engineering Science at the University of Johannesburg, South Africa. Prior to that, he was a faculty member at the University of Wisconsin, Milwaukee, United States of America. Jen received his Ph.D. in Mechanical and Aerospace Engineering from University of California, Los Angeles (UCLA), specializing in thermal aspects of grinding. He is currently leading the drive for application of atomic layer deposition equipment (first of its kind in Africa) for various applications (hydrogen, renewable energy, thin films, etc.). He is a fellow of the American Society of Mechanical Engineers (ASME), member of Academy of Science of South Africa (ASSAf), among others. Jen has written over 360 peer-reviewed articles, including 180 peer-reviewed journal papers, 16 book chapters, and five books to date.

1 Introduction

1.1 INTRODUCTION TO CORROSION

Corrosion can be defined in many ways. The word corrode is derived from the Latin "*corrodere,*" which means "to gnaw to pieces." The general definition of corrode is to wear away gradually. Therefore, corrosion can be defined as the destructive attack of a metal by an electrochemical or chemical reaction with its surrounding environment [1–3]. The environment consists of the entire surroundings in contact with the material. The primary factors to describe the environment are the following:

- Physical state – gas, liquid, or solid
- Chemical composition – constituents and concentration
- Temperature.

Non-metals are not included in this definition. The term corrosion in this book is confined to the chemical attack of metals. Plastic may swell or crack, wood decay or split, granite erode and Portland cement leach away. Metals and alloys employed in various industries are more vulnerable to corrosion, which cannot be avoided but can be managed in multiple ways. Corrosion has become a pressing issue because it not only destroys metals and alloys but also harms our resources and manual labor used in producing various metallic tools and products [4–6]. The most commonly used metals, such as iron, aluminum, steel, and copper, are more vulnerable to corrosion, leading to decreased use, life span, and property loss [7]. This corrosion renders turbine discs, pressure containers, boiler barrels, abrasives, hazardous chemical canisters, airplane components, automobile navigation devices, and superstructures like bridges unfit for use. Based on a review of the literature, over the last few decades, various cleaning procedures necessitate aggressive and corrosive media, including acid wiping, scraping, oil-well vacuuming, and acid pickling, which cause immense devastation to the outer layer of the metal [7–10], as depicted in Figure 1.1.

1.2 FORMS OF CORROSION

Corrosion occurs in different forms depending on how it establishes superficially on the metals or alloys. Corrosion appears in various forms depending on the metal

DOI: 10.1201/9781003441151-1

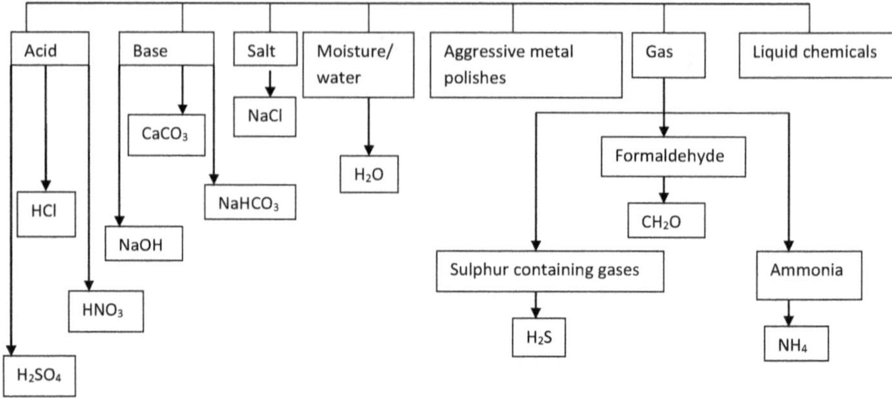

FIGURE 1.1 Causes of metallic corrosion.

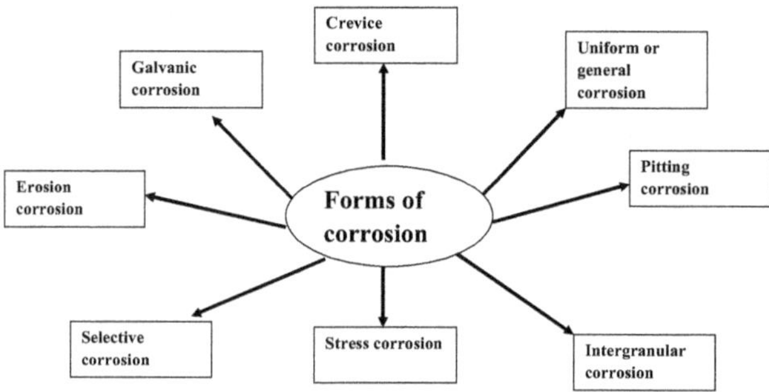

FIGURE 1.2 Forms of corrosion.

or alloy. Therefore, in terms of visual observation with either the naked eye or magnification, corrosion is readily classified into eight primary forms, as illustrated in Figure 1.2. The morphology of the attack is the basis for classification. Some of the most common forms of corrosion are schematically depicted in Figure 1.2. In theory, the eight forms of corrosion are distinct; in practice, however, there are several corrosion cases that fit into more than one category.

1.2.1 Uniform or General Corrosion

The most recognized form of corrosion is uniform corrosion. It is usually known as general corrosion. The attack, which generally occurs over the entire surface of the metal exposed to the corrosive medium, more or less uniformly, is called uniform corrosion. It almost always occurs in environments in which the rate of corrosion is comparatively low and well-controlled [11]. The most prone metals to uniform corrosion are low-alloy iron and magnesium alloys [11, 12]. The general attack is caused by local corrosion cell action; that is, multiple anodes and cathodes are active on the metal surface at any given time. Uniform corrosion over the bare metal surface progresses at almost the same rate. When exposed to open spaces, soils, and natural waters, cast iron and steel corrode uniformly, leading to the formation of rust. Collecting suitable materials, chemical-resistant protective coatings, cathodic and anodic protection, and corrosion inhibitors typically regulate uniform corrosion.

1.2.2 Galvanic Corrosion

In the electrolyte solution and an electron conductive path, galvanic corrosion, generally termed two-metal or bimetallic corrosion, is an electrochemical behavior of two dissimilar metals [13, 14],which arises if dissimilar metals or alloys with different electrode potentials are electrically connected in a standard electrolyte solution. The anode becomes more negative or an active metal, and the cathode of the galvanic corrosion cell becomes more positive or a noble metal. The less noble alloy typically demonstrates an increase in corrosion and a decrease or suppression of corrosion in the more noble material. The potential difference between two metals, the nature of the environment, the polarization behavior of metals, and the geometric relationship of the component metals all influence the degree of galvanic corrosion. Because of the direct relation of brass valves to the carbon steel pipe, or between copper tubing and steel pipe where the steel serves as the anode, and brass or copper as the cathode, the most common example of such corrosion behavior is typically found in process plant operations. To prevent galvanic corrosion, metal combinations with constituents that are as similar as possible in the respective galvanic series are selected. Additionally, measures such as seals, insulators, and coatings are used whenever possible to avoid direct contact between two different metals, controlling threaded junctions between materials widely separated in the galvanic series.

1.2.3 Pitting Corrosion

Pitting is the most devastating mode of corrosion attack which is often responsible for equipment failure in processing plants [15, 16], accounting for approximately 90% of metal destruction caused by corrosion. Corrosion pitting is a highly localized metal surface attack with no overall deterioration in the surrounding region. A pit may be defined as a cavity with a diameter equal to or less than the depth of a metal surface, which can cause the metal substrate to be penetrated rapidly. It is first identifiable as a white or gray powdery accumulation, comparable to dust, which smudges the surface. Tiny pits or holes could be detected on the surface as the coating is washed away. In some instances, pitting is developed over the overall metal surface, resulting in an irregular or rough surface profile. Pitting corrosion, as it is much harder to trace, prevent, and manage, is considered to be more dangerous than uniform corrosion damage. Depassivation of a small region is the driving force for pitting corrosion, which becomes anodic, whereas a huge area becomes cathodic, contributing to much more localized galvanic corrosion. Pitting corrosion can lead to a variety of metal/environment combinations. Pitting is also susceptible to engineering alloys, such as stainless steel and aluminum alloys, which form passive protective films on the surface. Corrosion by pitting is most likely to occur in solutions containing chloride, bromide, or hypochlorite ions [17]. Furthermore, in the absence of oxygen, an oxidizing cation causes pits to develop. As an alloying factor, molybdenum is advantageous, and stainless steel containing molybdenum, such as types 316, 317, 904, and 254 SMO, is more resistant than stainless steel not containing molybdenum. Conversely, metals susceptible to uniform corrosion may not experience pitting. Standard carbon steel corrodes evenly in seawater, whereas stainless steel is pitted. The maintenance of clean surfaces can manage corrosion by pitting, the application of protective coatings, and the use of immersion inhibitors or cathodic protection.

1.2.4 Crevice Corrosion

Crevice corrosion, which is deterioration in occluded areas, is one of the most damaging forms of localized material degradation [18, 19]. This type of corrosion is comparable to the pitting corrosion occurring in stagnant electrolytic conditions; it occurs due to the alteration of the local chemistry, that is, depletion of oxygen in the crevice, rise in values of pH with increased hydrogen ion concentration, and increase in chloride ion concentration. Oxygen depletion implies that, in the crevice region, cathodic reaction to oxygen reduction cannot be sustained, leading to metal dissolution. Crevice corrosion occurs on any metal and in any corrosive environment. However, metals such as aluminum and stainless steel that depend on their surface oxide film for corrosion resistance are particularly susceptible to crevice corrosion, especially in environments such as chloride ion-containing seawater. The substance responsible for the crevice corrosion formation need not be metallic. Crevice corrosion has been identified in materials such as wood, plastic, rubber, glass, and concrete. To prevent it, installations should be designed to allow complete drainage (avoiding corners and stagnant zones), use welds instead of bolted or riveted joints, employ only solid, nonporous seals, and utilize solid, nonabsorbent gaskets such as Teflon.

1.2.5 Intergranular Corrosion

Intergranular corrosion, called intercrystalline corrosion, is a localized preferential attack on the grain boundary or the areas around it [20, 21]. Little or no attack is observed on the main body of the grain. Lack of strength and flexibility results in this form of corrosion. Sometimes, the attack is quick, deeply penetrating the metal and causing failure. This form of corrosion results from the presence of impurities within limits, or local enrichment or degradation of one or more alloying elements. For instance, chromium corrosion in the grain boundary region occurs when austenitic stainless steels are sensitized by heating in the temperature range of approximately 500–800°C, leading to vulnerability to intergranular corrosion. The use of such low-carbon stainless steel grades or stabilized grades of stainless steel alloyed with strong carbide formers such as titanium or niobium could avoid these types of corrosion. These components combine with the carbon to form the appropriate carbides, thereby preventing chromium depletion.

1.2.6 Stress Corrosion Cracking

Structural components exposed to a combination of sustained tensile stress and a corrosive environment can fail prematurely at stress below the yield strength, a phenomenon known as stress corrosion cracking (SCC) [22]. Stress may occur due to loads being added, residual stresses from the production process (welding, heat treatment, machining, and grinding), or a combination of both. The group of commercial metals and alloys utterly resistant to SCC has not yet been identified. Standard austenitic stainless steels, such as AISI 304 and AISI 316, are typically vulnerable to SCC at temperatures above 60°C in Cl_2-containing environments. This form of corrosion could be prevented or reduced by removing residual stresses through heat treatments to alleviate stress, purifying the medium, choosing the most appropriate material, improving the surface quality, and applying external protection methods such as cathodic protection, inhibitors, and protective coatings.

1.2.7 Filiform Corrosion

It is practically a particular form of crevice corrosion, often called corrosion "under the film" [23, 24]. As moisture pervades the coating, such decay develops beneath painted or plated surfaces. Lacquers are most susceptible to the problem and "quick-dry" paints. Their use should be avoided until field experience demonstrates that they do not have any adverse effects.

1.2.8 Erosion Corrosion

The term "erosion" refers to decay due to mechanical forces. The assault is known as "erosion corrosion" when the factors leading to erosion increase the metal's corrosion rate. It is the product of an active chemical environment and high surface velocities of fluid, resulting from a fast fluid flow past a stationary object, such as the example of the oilfield check valve. Unlike surfaces from several other types of corrosion, surfaces that have suffered erosion corrosion usually are reasonably clean. Several metals and alloys, including those that rely on the formation of a surface oxide film for corrosion resistance, are susceptible to erosion–corrosion damage. However, softer metals such as Cu, Al, Pb alloys, and brass are naturally more vulnerable to corrosion by degradation compared to steel.

1.2.9 Selective Leaching or Dealloying

Selective leaching occurs when an element is displaced from an alloy through corrosion, leaving behind the elements that are much more resistant in that environment. For example, Zn is selectively leached out from the Cu–Zn alloy, leaving underneath a porous and brittle Cu-rich outer surface.

1.3 MECHANISM OF CORROSION

Metallic corrosion occurs only whenever the metals approach the corrosive agents directly [25]. The corrosion procedure is initiated by the materialization of a short-circuit electrochemical cell involving two different types of reactions at different terminals. Oxidation at an anode leads the metal atoms to lose their electrons, giving rise to the development of metal ions that tend to leave the metal surface. The other

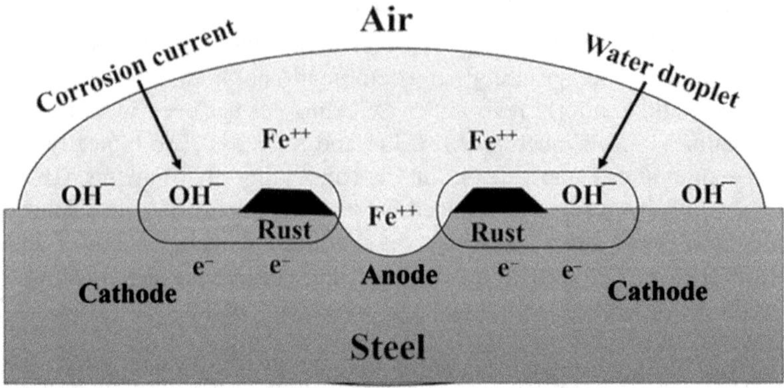

FIGURE 1.3 Mechanism of corrosion [26].

reduction reaction results in a gain of electrons taking place at the cathode. As a result of these reactions, metal atoms tend to lose their originality, properties such as permeability to fluids, mechanical strength, shape, size, and chemical state by changing into their corresponding metal ions. Rusting of iron is considered to explain the mechanism of corrosion, as depicted in Figure 1.3.

1.3.1 Anodic Reaction

Considering only metals and alloys, the anode is the site at which the metal is corroded, that is, at which metal dissolution occurs. The anodic reaction involves the oxidation of a metal to its ion because of the electric charge difference at the solid–liquid interface. A metal's generic anodic process can be described by a metal's oxidizing reaction to its ions, which passes into a solution:

$$M \rightarrow M^{+n} + ne^{-} \tag{1.1}$$

where n is the valence of metal, e indicates the electron, M is the generic metallic material, and M^{n+} is its ion, which passes into the solution.

1.3.2 Cathodic Reaction

The anodic dissolution reaction involves only the metallic phase, and the cathodic reaction involves the environment. The following reaction can represent the generic cathodic reaction:

$$R^{+} + e^{-} \rightarrow R^{o} \tag{1.2}$$

where R^{+} is a positive ion in solution, e^{-} is an electron in the metal, and R^{o} is the reduced chemical species. Several different cathodic reactions are possible and the environment determines the one that occurs. In an acidic aqueous environment and in the absence of dissolved oxygen, the primary cathodic reaction is hydrogen evolution:

$$2H^+ + 2e^- \rightarrow H_2 \tag{1.3}$$

While the same reaction in an alkaline aqueous solution occurs as:

$$2H_2O + 2e^- \rightarrow H_2 + 2OH^- \tag{1.4}$$

In the presence of dissolved oxygen, the cathodic reaction that is more thermodynamically favorable is oxygen reduction. In the acidic environment, it follows as under:

$$2O_2 + 4e^- + 4H^+ \rightarrow 4H_2O \tag{1.5}$$

In the alkaline environment, it proceeds as:

$$2O_2 + 4e^- + 2H_2O \rightarrow 4OH^- \tag{1.6}$$

Metal ion reduction and metal deposition are less common reactions and are most frequently found in chemical process streams.

$$M^{3+} + e^- \rightarrow M^{2+} \tag{1.7}$$

$$M^+ + e^- \rightarrow M \tag{1.8}$$

1.4 FACTORS INFLUENCING CORROSION

Electrochemical reactions that tend to occur whenever metals or alloys witness closeness with the atmosphere directly or indirectly are primarily responsible for corrosion to occur. Moreover, the properties of metals and alloys adversely impact the corrosion process. In addition to the causes mentioned in Figure 1.4, many environmental conditions affect corrosion including temperature, pH, and the nature of

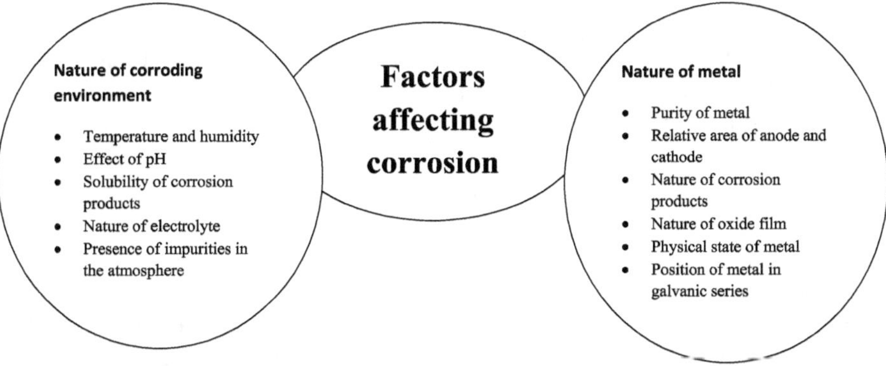

FIGURE 1.4 List of factors influencing corrosion.

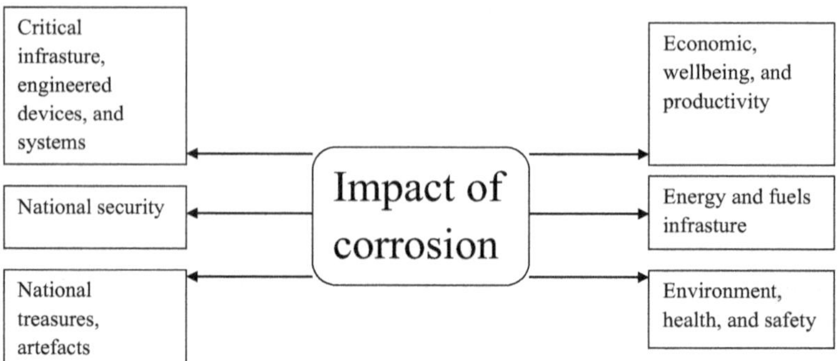

FIGURE 1.5 Impacts of corrosion.

metal [27–29]. The nature of metals readily stimulates the progression of corrosion. The metals with high reduction potential are low reactive and hence less susceptible to corrosion and vice versa. Corrosion reactions are electrochemical processes, usually intensified by increasing temperature, humidity, and conductance of the medium. However, an increase in concentration enhances the inhibition performance of most inhibitors; nevertheless, after a certain threshold, known as the "optimum concentration," further increase in concentration has negligible effect on the inhibitor molecule's inhibition characteristics. Some metals approaching a medium with low pH are more susceptible to corrosion than the metals in contact with a high pH medium. It has been noted that at pH > 10, the superficial speed of iron corrosion abruptly ends due to the formation of a protective hydrous oxide layer[7]. However, at pH = 3, at low air content, even in the absence of air, metals corrode due to the liberation of H_2 gas at the cathode terminal of the corrosion cell.

Numerous internal and external factors, as depicted in Figure 1.5 directly or indirectly influence the corrosion rate. Some of the factors that contribute to the rate of corrosion are outlined below.

1.4.1 Effect of the Electrical Conductance of Media

Because of the acceleration of the corrosion current, if the electrical conductance is strong, the electrical conductance of media plays a significant role in corrosion processes. The lower the medium's electric resistance R, the higher the corrosion current (*Icorr*) [30, 31]:

$$Icorr = \frac{V}{R} = \frac{(Ek - Ea)}{R} \tag{1.9}$$

All electrochemical (cathodic) safety parameters (shape selection, the density of the electrical current, etc.) depend on the electrical resistance of the media and the anode shape. The electrical resistance of the soil often depends on the distribution of stray electric currents. Electrochemical measurements rely, for example, on the electrical resistance of the media to determine the electrical current as a function of the electrical potential. The ohmic potential drop (IR drop) occurs during an electrical current flow between the working electrode and a reference electrode in the solution. There is a likely decrease due to the solution layer's electrical resistance. The higher a solution's electrical resistance, the greater the IR decrease. This error in potential measurements needs to be accounted for. An IR decrease may involve polarization measurements in pure water and nonaqueous solvents, and metal electrodes with coatings.

1.4.2 Influence of the pH on Metallic Corrosion

According to their relationship (corrosion resistance) to pH values, all metals can be divided into five groups:

1. The corrosion resistance of gold (Au), silver (Ag), platinum (Pt), palladium (Pd), rhodium (Rh), ruthenium (Ru), mercury (Hg), tantalum (Ta), niobium (Nb), osmium (Os), and iridium (Ir) is not affected by the pH value. At pH5–14, they are resistant to corrosion.
2. Iron (Fe), chromium (Cr), and manganese (Mn) are strongly corroded at temperatures above 80°C in acidic solutions (pH 4) and very hot alkaline solutions (pH 13.5). They often corrode but at neutral pH (6–8) at a low rate and are immune at pH 9–13.
3. Magnesium (Mg), titanium (Ti), hafnium (Hf), vanadium (V), and bismuth (Bi) are highly resistant to neutral and alkaline solutions (strong pH) and corrode at a high rate at low pH (in acid solutions).
4. In acid and neutral solutions, molybdenum (Mo), tungsten (W), and rhenium (Re) are resistant to corrosion, but they corrode in alkaline solutions.
5. Beryllium (Be), aluminum (Al), copper (Cu), zinc (Zn), cadmium (Cd), tin (Sn), lead (Pb), cobalt (Co), nickel (Ni), zirconium (Zr), gallium (Ga), and indium (In).

All metals that corrode in both acidic and alkaline liquids are called amphoteric metals. Only neutral solutions are resistant to these metals. It is clear that an area of pH where high resistance exists is distinct for various amphoteric metals. Aluminum is pH 4.5–8.3 resistant, and zinc is pH 6.5–12 resistant. The pH region of the high resistance of amphoteric metals will significantly change the temperature and presence of different ions in a solution. Amphoteric metals are often associated with iron and chromium as they dissolve at low and very high pH (above 13.5); the higher the temperature, the lower the pH value when the iron is dissolved in alkaline solutions.

1.4.3 Effect of Substances Forming Chemical Complexes with Metals on Corrosion

Substances of metals forming chemical complexes intensify their corrosion. Neutral molecules ammonia (NH_3) or ions (cyanides CN_2) can be these compounds. Ammonia is highly soluble in water, producing an ammonium hydroxide alkaline solution. In the presence of ammonia and dissolved oxygen in water, copper, and copper alloys are not immune due to the formation of a complex. Owing to their high corrosion resistance, we call gold and silver "noble" metals. Many people say they are immune to all media sources. But if you place them in an aqueous solution of sodium or potassium cyanide (NaCN or KCN), with these "noble" metals, anions of CN form chemical complexes. Hence, in solutions containing cyanide anions, gold and silver are not resistant.

1.4.4 Effect of Cations Participating in Cathodic Reactions

In addition to the presence of dissolved oxygen (O_2) in neutral and alkaline solutions and H_3O cations in acidic solutions, Fe^{3+} and Cu^{2+} cations may be involved in cathodic processes on the surface of the metal and accelerate corrosion. According to the reaction, Fe^{2+} cations formed in the cathodic reaction can be oxidized to Fe^{3+}. Therefore, they are very important to recognize and track such cations in a medium that can take part in cathodic reactions and accelerate corrosion.

1.4.5 Effect of Temperature

In chemistry, the temperature is the significant parameter for adjusting the rate of chemical reactions. Under significant variations in environmental conditions, metals and alloys are used. The thermodynamics and kinetics of metallic corrosion may be impacted by temperature. Temperature rises typically accelerate anodic and cathodic reactions, decrease gaseous oxygen dissolution in the media, and accelerate the diffusion of cathodic agents [32] such as O_2, H_3O, etc. The dissolution and transformation of corrosion products forming on metal surfaces are also altered by temperature. The effect of temperature is complex, and it is not easy to predict the behavior of metals with temperature changes. Temperature thus impacts corrosion by two factors: it accelerates both anodic and cathodic reactions and decreases the dissolved oxygen concentration (cathodic reaction as a result). These variables are identical at a specific temperature, and corrosion reaches the maximum value. Then the decrease in oxygen concentration in water prevails over the rise in the anodic and cathodic reaction rates at higher temperatures. Dissolved oxygen disappears from the "scene" as a cathodic partner, and corrosion decreases. In the breakdown of metals, oxygen is so critical that, in specific cases, it can accelerate corrosion and, in other cases, decrease and even stop it. In some instances, the dissolved oxygen concentration in the media is non-uniform and induces the development of differential aeration cells.

1.4.6 Effect of Dissolved Oxygen

In the corrosion of metals, dissolved oxygen plays a crucial and complex role. In neutral, alkaline, and acidic media, oxygen plays in cathodic procedures on the metal surface. Hence, for corrosion to occur, its existence is required. If dissolved oxygen is absent in water, corrosion in neutral and alkaline solutions decreases to almost zero. If the dissolved oxygen concentration increases, because of oxygen involvement in the cathodic processes, corrosion intensifies. Oxygen under some particular conditions (in the water of high purity) and high temperature may result in forming a passive protective dense film composed of metal oxides on the metal surface, and corrosion would decrease. One of the corrosion prevention strategies at power stations is the introduction of oxygen into the water. If the water contains ions, passive films may be damaged, and the passivation effect will not be obtained.

1.4.7 Effect of Dissolved Salts in Water on Corrosion

The chemical content of drinking water, cooling water, seas, oceans, rivers, and lakes varies. For all these types of water, their almost neutral pH (usually between 5.5 and 8.3), the presence of inorganic and organic compounds, and dissolved gases are the common denominators. Slight changes in the chemical content of water species can lead to drastic changes in metal corrosion rates. pH values from 5.5 to 8.3 do not affect the corrosion of metals. Metal cations formed in anodic metal dissolution react with hydroxide anions OH_2 formed in aqueous neutral solutions in a cathodic reduction of dissolved oxygen, and corrosion products are formed as hydroxides. The rate of corrosion depends on the form of salt and its concentration [33].

1.5 THE EFFECTS AND ECONOMIC IMPACT OF CORROSION

Corrosion poses a threat on the durability and strength of metals. Moreover, it affects the energy and manual efforts employed during the manufacture of various metallic tools and devices, indirectly marking the loss to any economy [34]. The economic factor is a fundamental motivation for much of the current research in corrosion. To reduce the economic impact of corrosion, corrosion scientists and engineers aim to reduce material losses that result from the corrosion of tanks, pipes, ships, bridges, machine components, etc. The loss of metal by corrosion is a waste of material, energy, water, and the human effort to produce and fabricate the metal structures. Economic losses are divided into direct and indirect losses. Indirect losses are more challenging to evaluate. An example includes shutdown, loss of product, loss of efficiency, contamination of product, overdesign, etc.

Direct losses include:

- The costs of replacing corroded structures and machinery or their components.
- Repainting structures where prevention of rusting is the prime objective.
- The capital costs and maintenance of cathodic protection systems for underground pipelines.

As per the latest research by the National Association of Corrosion Engineers (NACE), the total worldwide cost of corrosion, seems to be more than US$2.5 trillion, or nearly 4.2% of world GDP [35]. However, in South Africa, the overall expense of overall corrosion is assessed to be over R130 billion [36, 37]. Losses sustained by industry and governments amount to billions of dollars annually, approximately $276 billion in the United States, equivalent to 3.1% of the gross domestic product (GDP) [38]; for example, the sudden collapse because of corrosion fatigue of the Silver Bridge over the Ohio River at Point Pleasant, OH in 1967 resulted in the loss of 46 lives and cost millions of dollars. It has been estimated that 25–30% of this cost could have been avoided with proper corrosion preventive technology.

1.6 THE MAGNITUDE OF THE CORROSION ISSUE IN GENERAL

The cost of corrosion can be measured in social, environmental, and financial terms [39]. There have been several projects aimed at estimating the economic costs, but there needs to be more attention given to the social and environmental impacts. For example, a corrosion-induced oil pipeline failure in which the oil contaminates the surrounding ground and water is described in terms of the cleanup and repair costs but not in terms of the impact on the neighboring population, flora, and fauna, and the time for these to return to their former states. Fortunately, few corrosion-induced failures result directly in human deaths. Nevertheless, the resulting environmental contamination can cause death to countless other species and future sickness and suffering for the human population. The worst example of this was the Bhopal disaster of 1984 which resulted not only in the release of toxic gases and the immediate deaths of over 8,000 people but also caused longer-term health problems for a half-million people [40]. Another insidious form of corrosion is the long-term slow release of metallic ions, which can have unpredicted health impacts. According to George Hayes, Director General of the World Corrosion Organization [41]: The public hears about bridges and piers that collapse due to corrosion and environmental damage caused by pipeline failures, but does it ever hear about the loss of potable water from main water corrosion or ecological damage caused by corroded sewer lines? The answer is no. News media and government agencies need to find such events sufficiently newsworthy to alert the public. Yet, in many countries, the cost of water and wastewater system failures is far greater than any other sector of the economy.

Some consequences of corrosion are economic, and cause the following:

- Replacement of corroded equipment
- Overdesign to allow for corrosion
- Preventive maintenance, for example, painting
- The shutdown of equipment due to corrosion failure
- Contamination of products
- Loss of efficiency, such as when overdesign and corrosion products decrease the heat transfer rate in heat exchangers
- Loss of valuable product

Other consequences are social. These can involve the following issues:

- Safety, for example, sudden failure can cause fire, explosion, the release of a toxic product, and construction collapse

- Health, for example, pollution due to escaping effect from corroded equipment or due to a corrosion product itself
- Depletion of natural resources, including metals and the fuels used to manufacture them.

Although little attention has been paid to the social costs, the financial costs of corrosion have garnered a tremendous amount of attention in recent years, primarily to draw the attention of decision-makers to the magnitude of the problem. Corrosion Awareness Day highlights the immense costs associated with corrosion worldwide on April 24 each year. The primary drivers of the current research in the field of corrosion are the economic losses and environmental impact. To understand the economic effects of corrosion, several countries have undertaken studies on corrosion cost. The general conclusion is that corrosion consumes approximately 3% of the world's GDP ,approximately US$2.2 trillion [41]. The estimates of the financial cost to several nations are presented in Table 1.1, modified by Biezma and San Cristobal [42]. Uhlig published the first critical report on corrosion costs in 1949, where the annual corrosion cost in the United States was calculated to be 2.1% of the overall gross national product (GNP). Since then, several reports based on comprehensive studies have been published on the economic impacts of corrosion. A landmark study on the economic effects of corrosion undertaken in the United States in the late 1970s showed that the total loss due to corrosion in 1975 was $70 billion, approximately 5% of the GNP of that year. In 2002 another breakthrough report was published by the US Federal Highway Administration, which measured the direct costs associated with metallic corrosion in the US industrial sector. The NACE International initiated the research, and as part of the Transportation Equity Act for the 21st century, has a mandate from Congress. The overall annual direct corrosion cost was found to be $276 billion, about 3.1% of the GNP of the country. This amount contained only the direct costs associated with replacing defective materials and parts.

Indirect costs, such as output losses, effects on the environment, disturbances to transport, accidents, and fatalities, are calculated to be equal to direct costs. Other nations, like the United Kingdom, Japan, Australia, Kuwait, Germany, Finland, Sweden, India, and China, have conducted similar corrosion cost studies. The research ranged from casual and modest attempts to formal and comprehensive attempts. The general conclusion of these studies is that the annual cost of corrosion range from

TABLE 1.1
Estimated costs of corrosion in different countries [42]

Country	Year	Cost	% GNP	References
Japan	1997	3,938 B Yen	0.77	[43]
Australia	1983	US$2 B	1.5	[44]
Australia	2008	Aus. $1e$5 B	2	[39]
United States	2002	US$276 B	3.1	[45]
Canada	2003	CAD $ 41 B	–	[46]
Kuwait	1995	US$1 B	5.2	[47]

approximately 1–5% of each nation's GNP. Global economic losses due to corrosion, as estimated by NACE International in 2016, have been recorded in some publications at $2.5 trillion. In India, during 1984–1985, the direct cost of corrosion was calculated to be Rs. 40.76 billion, of which Rs. 18.04 billion was considered an avoidable cost if suitable corrosion preventive measures were in place. Another report on the cost of corrosion in India, published in 1997, estimated the annual losses due to corrosion to be Rs. 250 billion per year, which was 4% of GNP. As per the latest global study by NACE International, the cost to India's economy on account of corrosion is estimated to be 4.2% of the GDP. An outline of the country-wise contribution to the cost of corrosion, according to a report was published in 2013. According to this report, the combined cost of corrosion represents 3.4% of GDP. The various contribution is: the United States contributes 2.7%, India 4.2%, the European region 3.8%, the Arab world 5.0%, China 4.2%, Russia 4.0%, Japan 1.0%, Four Asian Tigers1Macau 2.5%, and rest of the world approximately 3.4% of the total global cost to corrosion. In addition to the economic consequences, corrosion results in structural failures that seriously impact human health and life and the surrounding environments. The safety and environmental concerns are extremely difficult to define in terms of cost. Therefore, corrosion problems must be addressed for safety and to minimize environmental pollution.

1.7 INTRODUCTION TO WASTE UTILIZATION

Waste is defined as portable objects that their owner has abandoned (subjective definition), or as requiring orderly disposal to protect the public welfare (objective definition). Figure 1.6 shows various types of waste. The coordinated development of resources, environment, energy, and population is a major social problem today. A lot of waste has not been effectively recycled, resulting in severe issues of environmental

FIGURE 1.6 Various types of wastes.

pollution and waste of resources. As a result, large-scale comprehensive utilization technologies have been initiated. To help support the effort, various governmental and non-governmental agencies established funding for comprehensive waste utilization technology. The utilization of waste has attracted worldwide attention. It is an effective way to improve resource utilization efficiency and productivity by using waste recycling methods. In the context of this book, the focus is on food and agricultural waste.

1.8 CHARACTERISTICS OF AGRICULTURAL AND FOOD WASTE

Whenever and wherever food, in any form, is handled, processed, packed, and stored, there will always be an unavoidable generation of waste. Waste is the most severe environmental problem in the manufacturing and processing of foods. Most of the volume of waste comes from cleaning operations at almost every stage of food processing and transportation operations. The quantity and general quality of this processing waste generated have several economic and environmental consequences concerning its disposal. The environmental repercussions of inadequately removing the pollutants from the waste stream can have ecological severe ramifications. Knowledge of the characteristics of food and agricultural waste is essential to developing economical and technically viable waste management systems that comply with current environmental policies and regulations. Management methods that may have been adequate with other industrial waste may only be feasible with food and agricultural waste if the methods are modified to reflect the characteristics of the waste and the opportunities it may hold. The waste produced in agricultural and food processing industry varies in both quantity and quality. These differences in quantity and quality dictate the type and capacity of waste management systems that should be deployed. A clear understanding of the characteristics of food and agricultural waste allows management decisions on treatment and utilization methods that are effective and economical; some food processing operations occur seasonally (processing of fruits and vegetables); this seasonality adds complexity to the waste management systems demanding substances in the waste of a high-intensity livestock farming operation are in the form of solid particulates.

1.9 HISTORY OF WASTE MANAGEMENT

As far back as 8000–9000 BCE, people learned to dispose of their waste outside their settlements. At that time, waste consisted of food scraps, mussel shells, bones, broken household items, and clay shards. It can be inferred that these people created dump sites to escape the nuisances of vermin, odor, and wild animals. In antiquity, many cities in Europe and Asia collected waste in clay containers and hauled it away. In many other areas, waste was gathered in, and feces were emptied and cleaned periodically. There are also records of regulations (Athens, 320 BCE) for the daily sweeping of streets by residents, although the relationship between hygiene and the scourges of humankind, such as the plague, smallpox, cholera, and the like, was not known at that time. In Athens, waste haulers were required to move the waste at least 2 km beyond the city wall.

Physicians like the Greek scholar Hippocrates (around 400 BCE), and the Arab Avicenna (Ibn Sina, 1000 CE), were the first to suspect the link between hygiene, contaminated water, spoiled food, and epidemics. The Roman Emperor Domitian (81–96 CE) ordered pest control as his advisors realized that a lack of cleanliness in the city was associated with increased rats, lice, bedbugs, etc. Emperor Vespasian (69–79 CE) had earthen urinals set up in public places and put a "urine tax" on the operators. By 300 CE, there were 144 public toilets in Rome, with running water underneath to carry away the waste. Nevertheless, numerous epidemics ravaged humankind, contributed to the decline of civilizations, and depopulated entire regions. Until the 19th century, roads, rivers, and groundwater were contaminated by human and animal waste. In the 6th and 14th centuries, epidemics ravaged densely populated Europe, claiming the lives of one-third of the population (25 million) in a very short period (1347–1352 CE) [48]. Not until the 15th century did city councils require the paving of streets so that no one would have to wade through feces and waste. Under the threat of stiff penalties, strict requirements for cleanliness were imposed upon the citizens. At that time, garbage cans were introduced, the streets were cleaned regularly, animal carcasses were collected, and the belongings of people who had died from the plague were burned.

Between 1850 and 1890 CE, a breakthrough in "waste management" occurred when the research of physicians and scientists like Ignaz Semmelweif, sanitation officer Thilenius, Louis Pasteur, and Robert Koch revealed bacteria and viruses as the causes of diseases. Their research demonstrated that the spread of diseases could be controlled by the presence or absence of public health measures. In the 19th century, 9 cholera epidemics claimed 380,000 lives between 1831 and 1873 in Prussia alone [48]. The link between public health measures and mortality was recognized and the scientific evidence's weight influenced the distractors. Demands for water and sewage treatment plants, airy apartments, careful food inspections, and preventive public health measures were being made and addressed to local officials. Engineers and technicians were now challenged to develop technologies to alleviate and solve these problems. Among other measures, this led to the construction of the first waste incinerators in England in 1876. The need for public health measures was dramatically highlighted in 1892 in the City of Hamburg when about 9,000 people fell victim to a cholera epidemic. The city had been pumping water contaminated with waste and feces from the Elbe River into the city's water supply system, encouraging the spread of the epidemic. The Prussian cities of Altona and Wandsbek were unaffected by the epidemic, although being located right next to the hardest hit parts of the City (St. Pauli and Eimsbiittel) because they had water treatment plants with sand filters. To this day, in the developing world, the direct relationship between environmental hygiene and life-threatening diseases is evident. Wherever waterways are used for waste disposal, where people are forced to economize on a living in garbage dumps, and where sanitary waste disposal, water, and sewage treatment are nonexistent because of natural catastrophes, wars, or under-development, even today, epidemics can spread virtually uncontrollably.

In October 1965, the Central Office for Waste Disposal was founded by the federal government and assigned to the Federal Health Office in Berlin. Experts and the discussions discussed problems thread bare, and issues of waste disposal were published in instructional pamphlets. These pamphlets served the local authorities responsible for waste disposal as early guidelines in their work. Consequent to the enormous economic growth in the postwar era, the waste generated by the 1970s had become a "waste avalanche." It was largely fueled by increased industrial production and private consumption. The production and consumption of goods from the extraction, production, finishing, and service sectors are distinguished by the fact that by-products and supplements are generated in addition to the product itself. The reasons for generating by-products are diverse and result from the complexities of industrial processing engineering. In recent years, the consumption of goods in private households has kept pace with increasing prosperity. As basic quantitative and qualitative human needs have largely been met, buying non-essential goods is playing an increasingly important role and has become one of the reasons for the rising volume of residential waste. The Department of Forestry, Fisheries, and Environment (DFFE) reported in 2022 that according to the State of the Waste Report, South Africa generates around 107.7 million tonnes of waste annually. Huge volumes of waste were landfilled, approximately 92.7% of hazardous waste, and 65% of general waste. This denotes a tremendous challenge to everyone involved in waste management, including private, industrial, and municipal waste generators; municipal and private waste management firms; consulting and design engineers; scientists; and legislators. The solutions to these tasks will include many exciting developments, especially for waste management. Efforts to minimize waste are increasingly visible in the industrial sector, induced and encouraged by pressure on the industry through legislation, rising disposal costs, declining disposal capacities, and simply for reasons of self-preservation. The goals of the Waste Act are as follows: Waste is to be avoided, Waste is to be recovered, and Waste must be disposed of in a manner that does not harm public welfare.

1.10 SUMMARY

Corrosion is a highly damaging and challenging phenomenon identified with huge safety threats and economic losses all over the world. According to the NACE, corrosion causes a loss of about 3.5% of the world's GDP. The cost of corrosion can be classified into direct and indirect. Corrosion can be devastating and lead to some unpleasant effects, particularly for industries that depend on non-corroded metal for operation. Its effects, which are not always immediately apparent, include reduced production, injuries, and significant financial losses. Industries lose billions of dollars out of their business every year, primarily due to corrosion, resulting in an enormous overall cost around the globe with this issue. However, through obtaining and using highly qualified corrosion practitioners, harmonizing standards, and continuing education, and training, all underpinned through fostering more excellent corrosion knowledge, the ability to reduce the cost by approximately $875 billion annually by effective implementation of current corrosion reduction technologies is easily achievable.

REFERENCES

1. Hossen, A., Mahmud, R. and Islam, A., 2023. Minimization of corrosion in aquatic environment–A review. *International Journal of Hydrology*, *7*(1), pp.9–16.
2. Olajire, A.A., 2017. Corrosion inhibition of offshore oil and gas production facilities using organic compound inhibitors – A review. *Journal of Molecular Liquids*, *248*, pp.775–808.
3. Sanni, O., Iwarere, S.A. and Daramola, M.O., 2023. Introduction: Corrosion basics and corrosion testing. In *Electrochemical and Analytical Techniques for Sustainable Corrosion Monitoring* (pp. 1–23). Elsevier.
4. Akpanyung, K.V. and Loto, R.T., 2019, December. Pitting corrosion evaluation: A review. *Journal of Physics: Conference Series*, *1378*(2), 022088.
5. Mustapaevich, D.K., Jaqsilikovna, J.G. and Axmeto'g'li, M.N., 2022. Corrosion of metals and factors affecting it. Methods of preventing corrosion of metals. *Academicia Globe: Inderscience Research*, *3*(9), pp.1–6.
6. Akpanyung, K.V. and Loto, R.T., 2019. Pitting corrosion evaluation: A review. *Journal of Physics: Conference Series*, *1378*(2), 022088.
7. Assad, H. and Kumar, A., 2021. Understanding functional group effect on corrosion inhibition efficiency of selected organic compounds. *Journal of Molecular Liquids*, *344*, 117755.
8. Verma, C., Thakur, A., Ganjoo, R., Sharma, S., Assad, H., Kumar, A., Quraishi, M.A. and Alfantazi, A., 2023. Coordination bonding and corrosion inhibition potential of nitrogen-rich heterocycles: Azoles and triazines as specific examples. *Coordination Chemistry Reviews*, *488*, 215177.
9. Han, T., Guo, J.X., Zhao, Q., Shen, T. and Zhang, S.L., 2022. Insights into the multiple applications of modified polyacrylamide as oilfield corrosion inhibitor and water phase tackifier. *Petroleum Science*, *19*(1), pp.397–408.
10. Haldhar, R., Asrafali, S.P., Periyasamy, T., Raorane, C.J., Vanaraj, R., Kim, S.C., Dagdag, O. and Ebenso, E.E., 2023. Corrosion monitoring of nanocomposites coatings. In *Smart Anticorrosive Materials* (pp.45–60). Elsevier.
11. Zehra, S., Mobin, M. and Aslam, J., 2022. An overview of the corrosion chemistry. In *Environmentally Sustainable Corrosion Inhibitors* (pp.3–23). Elsevier.
12. Xu, Q., Gao, K., Lv, W. and Pang, X., 2016. Effects of alloyed Cr and Cu on the corrosion behavior of low-alloy steel in a simulated groundwater solution. *Corrosion Science*, *102*, pp.114–124.
13. Francis, R., 2022. Galvanic corrosion. In *La Que's Handbook of Marine Corrosion* (pp.123–153). Wiley.
14. Lee, K., Sun, S., Lee, G., Yoon, G., Kim, D., Hwang, J., Jeong, H., Song, T. and Paik, U., 2021. Galvanic corrosion inhibition from aspect of bonding orbital theory in Cu/Ru barrier CMP. *Scientific Reports*, *11*(1), 21214.
15. Wang, R. and Shenoi, R.A., 2019. Experimental and numerical study on ultimate strength of steel tubular members with pitting corrosion damage. *Marine Structures*, *64*, pp.124–137.
16. Alamri, A.H., 2020. Localized corrosion and mitigation approach of steel materials used in oil and gas pipelines–an overview. *Engineering Failure Analysis*, *116*, 104735.
17. Wang, X., Qiao, L. and Zhang, L., 2022. Analysis of the causes of the cracking of tube bundles of 316L stainless steel Shell-and-Tube heat exchanger. *Engineering Failure Analysis*, *139*, 106484.

18. Ning, F., Wu, X. and Tan, J., 2019. Crevice corrosion behavior of Alloy 690 in high-temperature water. *Journal of Nuclear Materials*, *515*, pp.326–337.
19. Pan, C.Q., Zhong, Q.D., Yang, J., Frank Cheng, Y. and Chen, C., 2021. Investigating crevice corrosion of copper and copper alloys using wire beam electrode. *Corrosion Engineering, Science and Technology*, *56*(5), pp.407–418.
20. Zhang, X., Lv, Y., Hashimoto, T., Nilsson, J.O. and Zhou, X., 2021. Intergranular corrosion of AA6082 Al–Mg–Si alloy extrusion: The influence of trace Cu and grain boundary misorientation. *Journal of Alloys and Compounds*, *853*, 157228.
21. Sekhar, A.P., Samaddar, A., Mandal, A.B. and Das, D., 2021. Influence of Ageing on the intergranular corrosion of an Al–Mg–Si alloy. *Metals and Materials International*, *27*, pp.5059–5073.
22. Chen, L., Sheng, Y., Zhou, H., Li, Z., Wang, X. and Li, W., 2019. Influence of a MAO+ PLGA coating on biocorrosion and stress corrosion cracking behavior of a magnesium alloy in a physiological environment. *Corrosion Science*, *148*, pp.134–143.
23. Fayomi, O.S.I., Akande, I.G. and Odigie, S., 2019, December. Economic impact of corrosion in oil sectors and prevention: An overview. *Journal of Physics: Conference Series*, *1378*(2), 022037.
24. Kokilaramani, S., Al-Ansari, M.M., Rajasekar, A., Al-Khattaf, F.S., Hussain, A. and Govarthanan, M., 2021. Microbial influenced corrosion of processing industry by re-circulating waste water and its control measures– a review. *Chemosphere*, *265*, 129075.
25. Harsimran, S., Santosh, K. and Rakesh, K., 2021. Overview of corrosion and its control: A critical review. *Proceedings on Engineering Sciences*, *3*(1), pp.13–24.
26. Bazli, L., Yusuf, M., Farahani, A., Kiamarzi, M., Seyedhosseini, Z., Nezhadmansari, M., Aliasghari, M. and Iranpoor, M., 2020. Application of composite conducting polymers for improving the corrosion behavior of various substrates: A review. *Journal of Composites and Compounds*, *2*(5), pp.228–240.
27. Lou, Y., Chang, W., Cui, T., Wang, J., Qian, H., Ma, L., Hao, X. and Zhang, D., 2021. Microbiologically influenced corrosion inhibition mechanisms in corrosion protection: A review. *Bioelectrochemistry*, *141*, 107883.
28. Dastgerdi, A.A., Brenna, A., Ormellese, M., Pedeferri, M. and Bolzoni, F., 2019. Experimental design to study the influence of temperature, pH, and chloride concentration on the pitting and crevice corrosion of UNS S30403 stainless steel. *Corrosion Science*, *159*, 108160.
29. El Ibrahimi, B., El Mouaden, K., Jmiai, A., Baddouh, A., El Issami, S., Bazzi, L. and Hilali, M., 2019. Understanding the influence of solution's pH on the corrosion of tin in saline solution containing functional amino acids using electrochemical techniques and molecular modeling. *Surfaces and Interfaces*, *17*, 100343.
30. Chukwuike, V.I., Echem, O.G., Prabhakaran, S., AnandKumar, S. and Barik, R.C., 2021. Laser shock peening (LSP): Electrochemical and hydrodynamic investigation of corrosion protection pre-treatment for a copper surface in 3.5% NaCl medium. *Corrosion Science*, *179*, 109156.
31. Sanni, O., Ren, J. and Jen, T.C., 2022. Agro-industrial wastes as corrosion inhibitor for 2024-T3 aluminum alloy in hydrochloric acid medium. *Results in Engineering*, *16*, 100676.
32. Dou, Y., Han, S., Wang, L., Wang, X. and Cui, Z., 2020. Characterization of the passive properties of 254SMO stainless steel in simulated desulfurized flue gas condensates by electrochemical analysis, XPS and ToF-SIMS. *Corrosion Science*, *165*, 108405.

33. Pavlik, V., 1994. Corrosion of hardened cement paste by acetic and nitric acids part I: Calculation of corrosion depth. *Cement and Concrete Research, 24*(3), pp.551–562.
34. Koch, G., 2017. Cost of corrosion. In *Trends in Oil and Gas Corrosion Research and Technologies* (pp.3–30). Elsevier.
35. Verma, C., Quadri, T.W., Ebenso, E.E. and Quraishi, M.A., 2021. Polymer nanocomposites as industrially useful corrosion inhibitors: Recent developments. In *Handbook of Polymer Nanocomposites for Industrial Applications* (pp.419–435). Elsevier.
36. Verma, C., Ebenso, E.E. and Quraishi, M.A., 2017. Ionic liquids as green and sustainable corrosion inhibitors for metals and alloys: An overview. *Journal of Molecular Liquids, 233*, pp.403–414.
37. Assad, H. and Kumar, A., 2021. Understanding functional group effect on corrosion inhibition efficiency of selected organic compounds. *Journal of Molecular Liquids, 344*, 117755.
38. Aslam, R., Mobin, M., Zehra, S. and Aslam, J., 2022. A comprehensive review of corrosion inhibitors employed to mitigate stainless steel corrosion in different environments. *Journal of Molecular Liquids, 364*, 119992.
39. Hansson, C.M., 2023. An introduction to corrosion of engineering materials. In *Corrosion of Steel in Concrete Structures* (pp.1–16). Woodhead Publishing.
40. Gupta, P.K., 2004. Pesticide exposure– Indian scene. *Toxicology, 198* (1–3), pp.83–90.
41. Hays, G.F. and General, P.D., 2010. *World Corrosion Organization*. Corrodia NACE International.
42. Biezma, M.V. and San Cristobal, J.R., 2005. Methodology to study cost of corrosion. *Corrosion Engineering, Science and Technology, 40*(4), pp.344–352.
43. Committee on Cost of Corrosion in Japan, 2001. *Zair. to Kankyo/ Corrosion Engineering, 50*, pp. 490–512.
44. Cherry, B.W. and Skerry, B.S., 1983. Corrosion in Australia. The report of the Australian National Centre for Corrosion Prevention and Control Feasibility Study, Monash University
45. Koch, G.H., Brongers, M.P., Thompson, N.G., Virmani, Y.P. and Payer, J.H., 2002. *Corrosion Cost and Preventive Strategies in the United States* (No. FHWA-RD-01-156, R315-01). United States Federal Highway Administration.
46. Shipilov, S., 2009. What corrosion costs Canada: Or, can we afford to ignore corrosion. In *Materials Development and Performance of Sulphur Capture Plants* (pp.55–76). *Canadian Institute of Mining, Metallurgy, and Petroleum.*
47. Al-Kharafi, F.M., Al-Hashem, A. and Martrouk F. 1996. *Economic Effects of Metallic Corrosion in the State of Kuwait.* KISR Publications, Final Report No. 4761, December 1995.
48. Bilitewski, B., Härdtle, G., Marek, K., Bilitewski, B., Härdtle, G. and Marek, K., 1997. Introduction to waste management in Germany. In *Waste Management* (pp.1–20). Springer.

2 Corrosion Inhibitors
Fundamental Concepts

2.1 INTRODUCTION

The phenomenon of corrosion is highly damaging and challenging; therefore, various methods have been developed for its mitigation depending upon the nature of the metal and the surrounding environment. Given high economic and safety risks, corrosion inhibitors are one of the most effective and economical corrosion protection methods. Corrosion inhibitors can be defined as the chemicals or their mixtures that minimize or prevent corrosion. One of the oldest (before 1960) method was the use of inorganic compounds, including borates, molybdates, silicates, zinc salts, chromates, nitrates, etc. Inorganic inhibitors can be divided into two categories, depending on their significant influence on anodic or cathodic half-cell reactions. Most inorganic inhibitors become effective by forming a passive (oxide) film [1]. These are termed passivation or anodic-type corrosion inhibitors. However, some of them become effective by precipitating in the form of their oxides at the cathodic site. These are termed precipitating or cathodic-corrosion inhibitors. However, inorganic compounds are relatively expensive and toxic [2]; they were replaced by somewhat cheaper polyphosphonates, surface-activated chelates, carboxylates, phosphonates, gluconates, polyacrylates, and phosphonic acids.

Nevertheless, recent use of these chemicals is strictly restricted because of their toxicity. These highly toxic chemicals adversely affect aquatic and soil life after discharge. Furthermore, some inorganic compounds have bioaccumulation ability in living organisms, including humans. Because of the growing environmental awareness and strict environmental regulations, the use of the chemicals has been stopped, and relatively more eco-friendly organic corrosion inhibitors have replaced them. The use of organic compounds has been associated with economy, ecology, efficiency, and eco-friendliness. Eco-friendly corrosion inhibitors may be of natural or designed origin. Natural eco-friendly corrosion inhibitors include plant extracts, bio- and natural polymers, amino acids (AAs) and proteins, tannins, and vitamins. Organic compounds derived through proper design that avoids using toxic chemicals or solvents using energy-efficient heating are also regarded as eco-friendly corrosion inhibitors; these include compounds derived through solid-state reactions (SSRs) and multicomponent reactions (MCRs) [3].

DOI: 10.1201/9781003441151-2

Research on the use of environmentally friendly corrosion inhibitors and practices is gaining attention because of the high demands of "green chemistry" and growing ecological consciousness. The environmentally friendly properties of the corrosion inhibitors can be determined by assessing their toxicity to living organisms and to the environment, bioaccumulation, and biodegradability. Oslo Paris Commission (OSPAR) and Registration, Evaluation, Authorization, and Restriction (REACH) signify the international commissions that are used to appraise the above parameters. REACH represents a European Union regulation that was held on December 18, 2006, and was subjected to implementation on June 7, 2007 [4, 5]. OSPAR was subjected to implementation on September 22, 1992. There are several guiding principles and parameters to evaluate the effect of chemical substances in terms of their toxicity, biodegradability, and bioaccumulation. These guiding principles also help evaluate the impact of chemical substances on the environment and human health. The toxicity of a chemical can be determined using the effect of that chemical on a cell, tissue, or living organism. It is well established that the toxicity of any chemical depends upon gender, body weight, and age of the person; therefore, a population-level study is typically used to determine the toxicity of the chemical. Most chemical substances naturally undergo degradation or breakdown through microorganisms such as bacteria, fungi, and protozoa. These microorganisms are called decomposers. Generally, the process of natural decomposition is prolonged and takes several days, months, and sometimes years. Depending on the time required for decomposition, chemical substances can be divided into environmentally friendly or toxic.

2.2 GREEN CHEMISTRY PRINCIPLES AND GREEN CORROSION INHIBITION

In simple terms, green chemistry is the designing and modification of chemical processes that minimize or retard the implementation or production of toxic materials [6]. Green chemistry is also known as sustainable chemistry. In general, green chemistry involves a set of principles and covers the synthesis, design, use, and disposal of a chemical product, i.e. green chemistry. According to these principles, green chemistry:

- Minimizes environmental pollution at the molecular level.
- Includes all branches of chemistry, not a single discipline of chemistry.
- Reduces the adverse effects of processes and chemical products on human health and on the environment.
- Sometimes minimizes the discharge of toxic chemicals from previously existing products and processes.

It also includes modifying and designing existing products and processes to diminish their intrinsic hazards. For the green chemist, the sequence looks as follows:

- Designing chemical products that exert a less hazardous effect on the environment and human health.

- Making the feedstock of the solvents, chemicals, and reagents that have less hazardous effects on the environment and human health.
- Designing the processes and practices that avoid or minimize the formation of chemical waste.
- Designing the methods and procedures that prevent or reduce the use of energy or solvents.
- Use of chemical products derived from natural and renewable resources or collected (extracted) from the waste.
- Recycling and reuse of chemical products and chemicals.

Recently, corrosion scientists and engineers developed numerous environmentally friendly corrosion inhibition approaches. The attempts of green corrosion inhibition can be categorized as follows:

- Synthetic green corrosion inhibitors.
- Green corrosion inhibitors through proper designing.
- Green compounds through natural resources.
- Green corrosion inhibition practices.

2.2.1 Synthetic Green Corrosion Inhibitors

It is recalled that synthetic compounds stand for one of the most effective and economic classes of corrosion inhibitors. The cost-effectiveness of synthetic corrosion inhibitors is attributed to their ease of synthesis using inexpensive starting materials in high yield. However, not all synthetic approaches can be treated as equally important as some are associated with lower yield and the use of expensive starting chemicals, solvents, and catalysts. Primarily, compounds synthesized using multistep reactions (MRs) are associated with several drawbacks, including tedious, time-consuming, and low synthetic yield properties [7]. Because of their multistep nature, these reactions are associated with the numerous works-up and purification processes that finally reduce their yields. The MRs are also related to the colossal discharge of toxic chemicals, solvents, and catalysts into the surrounding environment due to the charge of toxic chemicals, solvents, and catalysts into the surrounding environment due to multiple purification steps. Consumption and discharge of expensive chemicals and solvents and adversely affecting the environment make the reaction highly costly.

Unlike this, MCRs that combine three or more reactant molecules in a single step have several advantages and emerged as one of the greenest synthetic protocols. The MCRs have the following benefits:

- Ease to perform due to one-step nature
- High synthetic yield, i.e., high atom economy due to a reduction in purification and works-up processes.
- Avoid or reduce the use of toxic and expensive solvents, as the majority of the MCRs are carried out in the solid phase.
- Inexpensive and time-saving.

- Reduced environmental pollution risks as MCRs are associated with minimum production and discharge of waste products.
- High selectivity.
- Facile automation.
- Ease purification.

Because of the association of MCRs with the aforementioned advantages, compounds derived from MCRs are regarded as sustainable [8]. It is essential to mention that several classes of organic compounds derived through MCRs are used as effective green corrosion inhibitors for different metals and alloys in various electrolytes. Corrosion inhibitors derived from energy-efficient microwave (MW) and ultrasound (US) irradiations represent another class of green corrosion inhibitors. The advantages of this nonconventional irradiation include:

- Rapid and uniform (homogeneous) heating and activation of reactant molecules.
- High yield (atom economy) and minimum waste side product formation.
- High selectivity and possibility of modification of selectivities.
- Simplification of purification and work-up procedures.
- Reduced environmental pollution as MW and US irradiations as connected with production of formation of waste products.
- Inexpensive and time-saving.
- Improved synthetic efficiency.

Evidently, because of their ease of handling and efficient heating, MW and US irradiations are time-saving and cost-effective [9]. Clark reported an 85-fold increase in the rate of Suzuki coupling reaction catalyzed by MW irradiation to conventional heating [10]. It is essential to mention that MW and US irradiations fulfill several requirements, including 12 principles of green chemistry. MW and US heating, in combination with MCRs, represent one of the greenest synthetic approaches. Apart from the compounds derived from MCRs with and without MW and US heating, solid-supported synthesis, mechanochemical mixing (MCM), and solid phase reactions represent some other common alternative synthetic routes for the production of green corrosion inhibitors. These syntheses are connected with lower waste generation and use and reduced use and discharge of chemicals and solvents.

2.2.2 Green Corrosion Inhibitors through Proper Designing

Numerous classes of compounds are environmentally friendly because of their properties. Ionic liquids (ILs) represent the best example of this type of compound [11]. ILs are salts in the liquid state, and they exist in a liquid state below 100°C. ILs are also called ionic melts, liquid electrolytes, ionic glasses, fused salts, ionic salts, or liquid fluids. Typically, ILs consist of organic cations and inorganic anions. Imidazolium, pyridinium, tetra-ammonium, and phosphonium represent the major cations in the ILs. Because of their lower vapor pressure, ILs are considered environmentally friendly alternatives meant for use in numerous industrial and biological applications, including corrosion inhibition. Owing to their fascinating properties such as nonhazardous nature, low volatility, nonflammability, and high thermal and chemical stabilities, ILs have been widely used for different applications. Further, the properties of ILs can be suitably tailored by appropriately varying the combination of cation and/or anion. Studies report that several classes of ILs are extensively used as green and effective corrosion inhibitors for various metals and their alloys in almost all types of electrolytes [12–22]. It has been observed that ILs show reasonably good corrosion inhibition efficiency. This is due to the presence of several electron-rich adsorption centers as well as their high solubility in the polar electrolytes. The protection efficiency of ILs can be appropriately enhanced for different metals and electrolyte systems by varying the nature of cations and anions. Polyethylene glycol (PEG) is another example of this type of compound, which is extensively used as a green corrosion inhibitor for different metals and alloys [23–30]. PEG is a well-established, inexpensive, and green chemical to be used for various industrial and biological applications.

2.2.3 Green Corrosion Inhibitors Derived through Natural Resources

Chemical species derived from natural resources are usually preferred to be used as environmentally friendly alternatives for several industrial and biological purposes. Chemical species derived from natural resources are widely used as corrosion inhibitors for different metals and alloys. One of the most common examples of natural corrosion inhibitors is a plant extract [31–33]. Because of their natural and biological origin, plant extracts are considered as environmentally friendly and suitable materials to be used for corrosion protection. Extracts derived through different parts of the plants, such as leaves, barks, fruits, flowers, and peels, are extensively used as corrosion inhibitors for different metal–electrolyte systems. Typically, each extract contains various phytochemicals that have electron-rich centers such as polar functional groups and aromatic ring(s) through which they adsorb effectively and exhibit metallic corrosion inhibition. It has also been observed that leaf extracts demonstrate better corrosion protection as compared to other extracts.

Another class of green corrosion inhibitors derived from natural resources is chemical medicines (drugs). It is essential to note that there are numerous drugs, especially ayurvedic medicines, derived from plants and other natural resources. Chemical medicines are commonly used as corrosion inhibitors because of their high protection effectiveness [34, 35]. The high inhibition efficiency of the drug molecules is attributed to their complex structures that acquire several electron-rich centers, including polar functional groups such as –OH, –NO, –COOH, –CONH, –COCI, and –COOC, H. and aromatic ring(s) through which they can easily get adsorbed. The use of medicines derived from natural resources should be encouraged; however, the implementation of synthetic medicines cannot be treated as environmentally friendly. In general, the synthesis of medicines requires an extremely high level of work-up and ultra-purification, making their synthesis expensive. Further, the synthesis of chemical medicines requires MRs that use and discharge a huge amount of toxic solvents and chemicals into the surrounding environment and pollute it.

Carbohydrates, AAs, and their derivatives are also extensively used as corrosion inhibitors for different metal–electrolyte combinations [36–40]. Owing to their natural and biological origin, carbohydrates, AAs, and their derivatives are treated as environmentally friendly alternatives for corrosion inhibitors. Despite their complex molecular structures, carbohydrates and their derivatives are highly soluble in polar electrolytes. This is attributed to their polar functional groups such as OH, –CONH, and –NH. Due to their complex structure, carbohydrates and their derivatives provide excellent metallic surface coverage in the form of their high inhibition efficiency. A literature study showed chitosan and cellulose are the most frequently used carbohydrate-based polymers that are most extensively used as corrosion inhibitors. Along with polysaccharides, mono- and oligo-saccharides have also been used extensively as corrosion inhibitors. Similarly, AAs and their derivatives, including heterocyclic compounds and proteins, are extensively used as corrosion inhibitors in different metal–electrolyte systems. Likewise, oleochemicals, which are chemicals derived from natural oils and fats, can also be treated as natural resources to be used as corrosion inhibitors. Typically, these chemicals possess polar functional groups (hydrophilic) through which they easily get adsorbed and their nonpolar hydrophobic chains) act as water repellent, i.e., oleochemicals behave as surfactants. A literature study shows that several reports have been published in which oleochemicals or their derivatives are used as corrosion inhibitors [41, 42].

2.3 GREEN CORROSION INHIBITION PRACTICES

Currently, computational modeling has emerged as an environmentally safe practice for designing effective corrosion inhibitors. The greenness of these models is because various reactivity information as well as reactivity parameters of a molecule can be determined before their synthesis. Generally, the synthesis of organic compounds is always an expensive and non-environmentally friendly approach. Consequently, computational simulations that provide functional reactivity parameters are gaining particular attention. These simulations provide several parameters to which relative chemical reactivity, including the corrosion inhibition effect, can be suitably delineated. It is well established that metal–inhibitor interaction involves charge (electrons) sharing. The molecular sites responsible for electron donation and acceptance can be derived through density functional theory [43, 44]. Evidently, in the frontier molecular orbitals, the highest occupied molecular orbital (HOMO) is responsible for electron donation, whereas the lowest unoccupied molecular orbital (LUMO) is responsible for electron acceptance. Molecular dynamics (MD) and Monte Carlo (MC) simulations are common computational models through which the orientation of the inhibitor molecules on a metallic surface can be determined. Generally, an inhibitor molecule with a flat or horizontal orientation would cover a larger metallic surface, and it would behave as a better corrosion inhibitor as compared to a compound having a vertical orientation. Along with orientation, MD and MC simulations offer information on whether the interaction between inhibitor molecules is spontaneous.

2.4 CLASSIFICATION OF CORROSION INHIBITORS

Corrosion inhibitors are chemical species that inhibit or decrease the corrosion rate in their presence at low concentrations. Corrosion inhibitors can be divided into various classes according to their nature, source of origin, mechanism of inhibition, effect on the environment, and mode of adsorption on the metallic surface. In a broader sense, corrosion inhibitors may be classified as organic and inorganic types. They can be further divided into different subclasses depending on various aspects. A broad classification of corrosion inhibitors is presented in Figures 2.1 and 2.2. Some main organic corrosion inhibitors are heterocyclics, different families of organic compounds, polymers, and their composites, oligomers and their composites, and nanomaterials. Major inorganic corrosion inhibitors are metal oxides and their organic and inorganic mixed composites.

2.4.1 Classification of Inorganic Corrosion Inhibitors

Inorganic corrosion inhibitors can be divided into two classes, namely anodic- and cathodic-type depending upon their inhibitive influence on either of the corrosion reactions. Inorganic compounds that primarily influence the rate of anodic reactions are called anodic corrosion inhibitors. Generally, these compounds become effective by forming the surface protective oxide film (passive film). The process of formation of an oxide film is generally referred to as passive film or passive layer, known as passivation, and such inhibitors are known as passivating inhibitors or simply passivators. Passivators may be further classified into oxidizing anions and un-oxidizing anions. Chromate and nitrite passivate the metallic surface even in the absence of oxygen and they are called oxidizing anions. However, tungstate, un-oxidizing anions, chromate, and nitrite passivate the metallic surface even in the absence of oxygen and they are termed oxidizing anions. Conversely, tungstate, phosphate, and molybdate

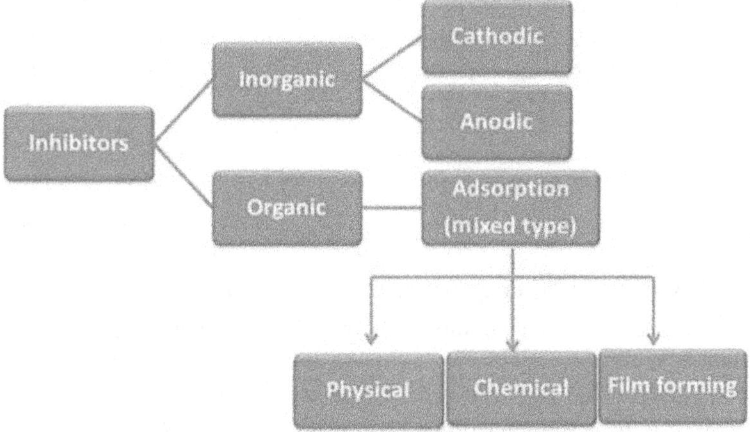

FIGURE 2.1 Illustration of classifications of corrosion inhibitors [5].

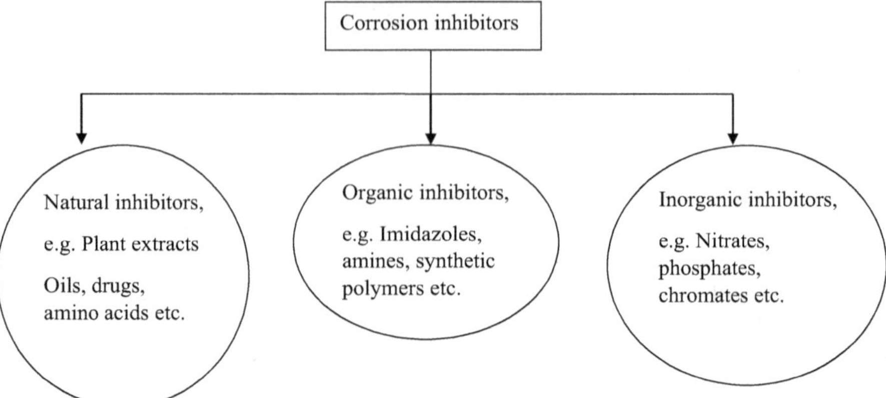

FIGURE 2.2 Classification of corrosion inhibitors based on their inorganic and organic nature.

passivate the metallic surface only in the presence of oxygen; these anions are called nonoxidizing anions.

Cathodic corrosion inhibitors can be divided into three classes: cathodic poisons, cathodic precipitates, and scavengers. Chemical species that slow the rate of cathodic reactions are known as cathodic poisons. For example, antimony and arsenic make the connection of hydrogen more complex and slow down the cathodic hydrogen evolution reaction. Some of the inorganic elements precipitate on cathodic sites in the form of their oxides and retard or slow down the rate of cathodic reactions. These elements are termed cathodic precipitates. For instance, magnesium, calcium, and zinc precipitate in the form of their oxides and function as cathodic precipitates. Generally, these precipitates avoid the diffusion of corrosive species, such as aggressive ions like chloride and sulfate. It has been observed that the rate of cathodic reaction is faster in the presence of oxygen; therefore, cathodic reactions can be controlled by removing oxygen. Some chemical species, such as methyl sulfate and hydrazine, are used to remove the oxygen from the corrosive environments. These chemicals are known as oxygen scavengers.

2.4.2 Classification of Organic Corrosion Inhibitors

Some of the major classifications of organic corrosion inhibitors are described below.

2.4.2.1 Classification of Organic Compounds Based on the Origin

Based on the origin, organic compounds may be classified as synthetic and natural corrosion inhibitors. Synthetic corrosion inhibitors are synthesized in laboratories employing different synthetic methods. Synthetic corrosion inhibitors include heterocyclics, nanomaterials, composites and nanocomposites, polymers and polymers composites, ILs, different families of organic compounds, PEG, etc. Natural corrosion inhibitors are derived from natural resources. Compounds derived from natural resources may be used with and without purification. Examples of natural corrosion inhibitors include plant extracts, natural medicines, carbohydrates, AAs, oleochemicals, natural and bio-polymers, natural gums, latex, etc.

2.4.2.2 Classification of Organic Compounds Based on Their Nature

Based on their nature, organic corrosion inhibitors can be divided into aliphatic and aromatic types. Aliphatic corrosion inhibitors can be further classified as acyclic and alicyclic types. Acyclic compounds are those that do not contain cyclic ring(s) in their molecular structures. One of the most common examples of this category is amines with a polar amino (–NH) functional group and a nonpolar hydrophobic hydrocarbon chain. Other examples include surfactants, polymers, alcohols, ethers, etc. Acyclic corrosion inhibitors are compounds that have one or more cyclic ring(s) in their molecular structures. Examples of this category of compounds include acyclic amines, cycloalkanes, cycloalkenes, cycloalkynes, and their derivatives. Reports on the corrosion inhibition effect of acyclic and alicyclic compounds are relatively limited because of their low inhibition effect. However, the corrosion inhibition effect of aromatic compounds, especially for hetero-aromatic compounds, is widely studied. Benzene, aniline, phenol, naphthalene, anthracene, and their derivatives are reported as effective corrosion inhibitors for different metals and alloys in various electrolytes. Examples of hetero-aromatic compounds include pyridine, quinoline, isoquinoline, and their derivatives. In a number of reports, the corrosion inhibition effect of the heterocyclic compounds such as pyrrole, furan, thiophene, and their derivatives has also been reported. Hetero-aromatic compounds react with the metallic surface using their nonbonding electrons of the heteroatoms (O, N, S, and P) and π-electrons of the multiple bonds. Typically, the corrosion inhibition effect of these compounds enhances with increasing conjugation in the form of polar functional groups and multiple bonds of the side chain(s).

2.4.2.3 Classification of Organic Compounds Based on Inhibition Mechanism

Organic corrosion inhibitors can also be classified based on their mode of action. Some of the compounds primarily inhibit anodic Tafel reactions. These compounds are known as anodic corrosion inhibitors [45]. Cathodic corrosion inhibitors are those that principally inhibit cathodic Tafel reaction. However, earlier study shows that most of the organic compounds function as mixed-type corrosion inhibitors, i.e., these compounds adversely affect anodic as well as cathodic reactions [46].

2.4.2.4 Classification of Organic Compounds Based on Molecular Size

Organic corrosion inhibitors can also be divided into different classes depending on their molecular size. Based on their molecular size, the main class of organic compounds would be molecular, macromolecular, and nanosized organic corrosion inhibitors. Examples of molecular corrosion inhibitors include different families of organic compounds. Macromolecular corrosion inhibitors can be categorized further into oligomeric and polymeric corrosion inhibitors. Literature study shows that natural and synthetic oligomers and polymers are extensively studied as corrosion inhibitors [47–49]. Chitosan, chitin, cellulose, natural gums, proteins, and their derivatives are some well-known natural polymeric corrosion inhibitors. Synthetic polymers and oligomers are also reported as effective corrosion inhibitors. The high inhibition efficiency of the macromolecular compounds is based on their highly large molecular sizes that provide excellent surface coverage. Among the nanosized organic corrosion inhibitors, graphene (G), graphene oxide (GO), carbon dots (CDs), carbon nanotubes (CNTs), and their derivatives and composites are extensively used. Generally, these nanomaterials show reasonably good anti-corrosion effects at relatively low concentrations, which can be attributed to their high surface areas as nanomaterials have extremely high surface area. Owing to their high surface areas, nanomaterials provide excellent surface coverage and function as effective corrosion inhibitors.

2.4.2.5 Classification of Organic Compounds Based on Their Effect on the Environment

Organic compounds may or may not have adverse effects on the surrounding environment. Accordingly, corrosion inhibitors can be divided into two classes: toxic- and non-toxic or environmentally friendly corrosion inhibitors. Most of the synthetic corrosion inhibitors are toxic as they use huge amounts of harmful chemicals and solvents and expensive catalysts for their syntheses. However, synthetic corrosion inhibitors derived through MCRs with and without ultrasound and microwave irradiations, mechanochemical mixing and solid-phase reaction, and solid-supported synthesis can be treated as environmentally friendly because these methods offer several advantages including high synthetic yield, ease of performance, production of less waste products, time-saving, less tedious, and many more. Further, compounds synthesized in environmentally friendly solvents, such as water and supercritical CO, can also be assumed as environmentally friendly in nature.

2.4.2.6 Classification of Organic Compounds Based on State of Application

Organic corrosion inhibitors can also be divided into two classes, based on their mode of application: aqueous phase corrosion inhibitors and coating phase corrosion inhibitors. Generally, coating phase corrosion inhibitors are used to provide long-term protection and are employed along with some polymer supports, such as epoxy resins mostly in coating and painting. However, aqueous-phase corrosion inhibitors are used to protect metallic corrosion in aqueous electrolytes. It is essential to note that only water-soluble compounds can be used as aqueous phase corrosion inhibitors.

2.4.2.7 Classification of Organic Compounds Based on Their Mode of Adsorption

Based on the mode of adsorption, organic compounds may be physical, chemical, and physiochemical (mixed) types. Generally, charged inhibitor molecules interact with the charged metallic surface through the electrostatic force of attraction. Physical adsorption generally is consistent with the value of the standard Gibb's free energy (AG) of =20 kJ/mol or more positive [50]. It is important to note that the extent of physical adsorption decreases with increasing temperature. Chemical corrosion inhibitors interact with metallic surfaces using a charge-sharing mechanism. In general, chemical adsorption is consistent with the standard Gibb's free energy value of –40 kJ/mol or more negative [51]. Literature study reveals that the adsorption of most of the organic compounds follows a physiochemisorption mechanism for which values of standard Gibb's free energy (AG) range between 20 kJ/mol and –40 kJ /mol.

2.5 TOXICITY OF CORROSION INHIBITORS

The use of corrosion inhibitors in different industries for corrosion control is a well-known practical technique, especially in the oil and gas industry. However, some of the corrosion inhibitors that are still in use are regarded as toxic and harmful to the environment, such as those based on nitrate and chromate. They were used in the treatment of industrial water, that is, in the recirculation of cooling water. Unfortunately, both are highly toxic. Thiourea derivatives form an essential class of corrosion inhibitors used in the past; however, they enhance hydrogen uptake by metals as well as isomerize into toxic compounds; thus they cannot be safely used. Another example is corrosion protection by n-alkanthiols on iron and copper. Despite the effective performance exhibited by the n-alkanethiols, their widescale applicability is not acceptable due to their toxic nature. Similarly, another class is acetylene alcohols, which are applied as acidizing inhibitors. They are commercially viable and effective, but they are applicable only at very high concentrations. They are highly toxic and cause problems in the handling and disposal of the waste. Several chemical compounds that are still used for the inhibition of corrosion are known to be extremely toxic. Addressing this issue and identifying inhibitors that effectively control corrosion, while being nonhazardous, remains an important subject for researchers. The environment friendliness of organic compounds has opened up numerous opportunities to replace the existing corrosion inhibitors that are highly toxic.

2.6 SUMMARY

Corrosion inhibitors are chemical species that inhibit or decrease the corrosion rate at low concentrations. They can be classified into different classes and categories, primarily as organic- and inorganic-type corrosion inhibitors. Inorganic inhibitors can be further classified into anodic- and cathodic-type inorganic inhibitors. Some inorganic compounds, known as oxidizing anions, passivate the metallic surface in the absence of oxygen. However, some anions, known as un-oxidizing anions, require oxygen to passivate the metallic surface. Organic compounds can also be classified as natural or synthetic types based on there origin. Based on their major corrosion inhibition effect, organic corrosion inhibitors may be classified as anodic-, cathodic, or mixed-type. According to their mode of adsorption, these organic corrosion inhibitors can be classified as physical, chemical, or physiochemical types. Organic corrosion inhibitors can be classified into molecular, nanosized, and macromolecular types based on their molecular size. Based on their environmental effect, organic corrosion inhibitors can be divided into environmentally friendly or environmentally toxic types.

Due to growing ecological awareness and strict environmental regulations, traditional corrosion inhibitors are highly restricted. Recently, corrosion scientists and engineers have been trying to develop green corrosion inhibitors and green corrosion inhibition practices. Chemical species derived from raw natural materials can also be considered as green alternatives for use as corrosion inhibitors. Compounds derived through MCRs with and without MW and US irradiations, MCMs, solid-phase reactions, solid-supported reactions (SSRs), etc. represent the major synthetic green corrosion inhibitors. Furthermore, compounds synthesized in green solvents such as ILs, water, and supercritical CO_2 can also be considered green. Along with green corrosion inhibitors, corrosion monitoring with computational simulations, including density functional theory and MD and MC simulations, can be deemed environmentally friendly approaches to corrosion mitigation because they do not involve the use of any toxic chemicals and expensive instruments.

REFERENCES

1. Pradhan, B., 2022. A study on effectiveness of inorganic and organic corrosion inhibitors on rebar corrosion in concrete: A review. *Materials Today: Proceedings.*
2. Nezhad, A.N., Davoodi, A., Zahrani, E.M. and Arefinia, R., 2020. The effects of an inorganic corrosion inhibitor on the electrochemical behavior of superhydrophobic micro-nano structured Ni films in 3.5% NaCl solution. *Surface and Coatings Technology*, *395*, 125946.
3. Verma, C., Alrefaee, S.H., Quraishi, M.A., Ebenso, E.E. and Hussain, C.M., 2021. Recent developments in sustainable corrosion inhibition using ionic liquids: A review. *Journal of Molecular Liquids*, *321*, 114484.
4. Klingel, T., Kremer, J.I., Gottstein, V., Rajcic de Rezende, T., Schwarz, S. and Lachenmeier, D.W., 2020. A review of coffee by-products including leaf, flower, cherry, husk, silver skin, and spent grounds as novel foods within the European Union. *Foods*, *9*(5), p.665.
5. Rüdel, H., Körner, W., Letzel, T., Neumann, M., Nödler, K. and Reemtsma, T., 2020. Persistent, mobile and toxic substances in the environment: A spotlight on current research and regulatory activities. *Environmental Sciences Europe*, *32*(1), pp.1–11.
6. Dariva, C.G. and Galio, A.F., 2014. Corrosion inhibitors–principles, mechanisms and applications. *Developments in Corrosion Protection*, *16*, pp.365–378.
7. Verma, C., 2021. *Handbook of Science &Engineering of Green Corrosion Inhibitors: Modern Theory, Fundamentals &Practical Applications*. Elsevier.
8. Abou-Shehada, S., Mampuys, P., Maes, B.U.W., Clark, J.H. and Summerton, L., 2017. An evaluation of credentials of a multicomponent reaction for the synthesis of isothioureas through the use of a holistic CHEM21 green metrics toolkit. *Green Chemistry*, *19*(1), pp.249–258.
9. Yang, H., Zhang, Q., Chen, Y., He, Y., Yang, F. and Lu, Z., 2018. Microwave–ultrasonic synergistically assisted synthesis of ZnO coated cotton fabrics with an enhanced antibacterial activity and stability. *ACS Applied Bio Materials*, *1*(2), pp.340–346.
10. Kokel, A., Schäfer, C. and Török, B., 2017. Application of microwave-assisted heterogeneous catalysis in sustainable synthesis design. *Green Chemistry*, *19*(16), pp.3729–3751.

11. Kumar, S.S. and Kumar, S.R., 2021. Ionic liquids as environmental friendly cutting fluids – a review. *Materials Today: Proceedings*, *37*, pp.2121–2125.
12. Verma, C., Ebenso, E.E. and Quraishi, M.A., 2017. Ionic liquids as green and sustainable corrosion inhibitors for metals and alloys: An overview. *Journal of Molecular Liquids*, *233*, pp.403–414.
13. Kobzar, Y.L. and Fatyeyeva, K., 2021. Ionic liquids as green and sustainable steel corrosion inhibitors: Recent developments. *Chemical Engineering Journal*, *425*, 131480.
14. Zeng, X., Zheng, X., Guo, L., Xu, Q., Huang, H. and Tan, B., 2021. Three imidazole ionic liquids as green and eco-friendly corrosion inhibitors for mild steel in sulfuric acid medium. *Journal of Molecular Liquids*, *324*, 115063.
15. Zaky, M.T., Nessim, M.I. and Deyab, M.A., 2019. Synthesis of new ionic liquids based on dicationic imidazolium and their anti-corrosion performances. *Journal of Molecular Liquids*, *290*, 111230.
16. Cao, S., Liu, D., Zhang, P., Yang, L., Yang, P., Lu, H. and Gui, J., 2017. Green Brönsted acid ionic liquids as novel corrosion inhibitors for carbon steel in acidic medium. *Scientific Reports*, *7*(1), p.8773.
17. Hu, P., Wu, Z., Wang, J., Huang, Y., Liu, Q. and Zhou, S.F., 2020. Corrosion inhibiting performance and mechanism of protic ionic liquids as green brass inhibitors in nitric acid. *Green Energy & Environment*, *5*(2), pp.214–222.
18. Qiang, Y., Zhang, S., Guo, L., Zheng, X., Xiang, B. and Chen, S., 2017. Experimental and theoretical studies of four allyl imidazolium-based ionic liquids as green inhibitors for copper corrosion in sulfuric acid. *Corrosion Science*, *119*, pp.68–78.
19. Haldhar, R., Raorane, C.J., Mishra, V.K., Periyasamy, T., Berisha, A. and Kim, S.C., 2023. Development of different chain lengths ionic liquids as green corrosion inhibitors for oil and gas industries: Experimental and theoretical investigations. *Journal of Molecular Liquids*, *372*, 121168.
20. Verma, C., Alrefaee, S.H., Quraishi, M.A., Ebenso, E.E. and Hussain, C.M., 2021. Recent developments in sustainable corrosion inhibition using ionic liquids: A review. *Journal of Molecular Liquids*, *321*, 114484.
21. Aslam, R., Mobin, M., Obot, I.B. and Alamri, A.H., 2020. Ionic liquids derived from α-amino acid ester salts as potent green corrosion inhibitors for mild steel in 1M HCl. *Journal of Molecular Liquids*, *318*, 113982.
22. Ardakani, E.K., Kowsari, E., Ehsani, A. and Ramakrishna, S., 2021. Performance of all ionic liquids as the eco-friendly and sustainable compounds in inhibiting corrosion in various media: A comprehensive review. *Microchemical Journal*, *165*, 106049.
23. Umoren, S.A., Ogbobe, O., Okafor, P.C. and Ebenso, E.E., 2007. Polyethylene glycol and polyvinyl alcohol as corrosion inhibitors for aluminium in acidic medium. *Journal of Applied Polymer Science*, *105*(6), pp.3363–3370.
24. Abdallah, M., Megahed, H.E., Radwan, M.A. and Abdfattah, E., 2012. Polyethylene glycol compounds as corrosion inhibitors for aluminum in 0.5 M hydrochloric acid solution. *Journal of American Science*, *8*(1), pp.49–55.
25. Boudellioua, H., Hamlaoui, Y., Tifouti, L. and Pedraza, F., 2019. Effects of polyethylene glycol (PEG) on the corrosion inhibition of mild steel by cerium nitrate in chloride solution. *Applied Surface Science*, *473*, pp.449–460.
26. Abd-Elaal, A.A., Shaban, S.M. and Tawfik, S.M., 2017. Three Gemini cationic surfactants based on polyethylene glycol as effective corrosion inhibitor for mild steel in acidic environment. *Journal of the Association of Arab Universities for Basic and Applied Sciences*, *24*, pp.54–65.

27. Ashassi-Sorkhabi, H. and Kazempour, A., 2020. Influence of fluid flow on the performance of polyethylene glycol as a green corrosion inhibitor. *Journal of Adhesion Science and Technology*, *34*(15), pp.1653–1663.
28. Verma, C. and Hussain, C.M., 2022. Polyethylene glycol and its derivatives as environmental sustainable corrosion inhibitors: A literature survey. In *Environmentally Sustainable Corrosion Inhibitors* (pp.321–333). Elsevier.
29. Talat, N.T., Dahadha, A.A., Abunuwar, M. and Hussien, A.A., 2023. Polyethylene glycol and polyvinylpyrrolidone: potential green corrosion inhibitors for copper in H_2SO_4 solutions. *International Journal of Corrosion and Scale Inhibition*, *12*(1), pp.215–243.
30. Izionworu, V.O., Byadi, S., Arukalam, I.O. and Nik, W.M.N.W., 2022. Inhibitive mechanism of acid corrosion of cold-rolled steel by polyethylene glycol via experimental, computational and surface analytical approaches. *Journal of Bio-and Tribo-Corrosion*, *8*, pp.1–14.
31. Alrefaee, S.H., Rhee, K.Y., Verma, C., Quraishi, M.A. and Ebenso, E.E., 2021. Challenges and advantages of using plant extract as inhibitors in modern corrosion inhibition systems: Recent advancements. *Journal of Molecular Liquids*, *321*, 114666.
32. Miralrio, A. and Espinoza Vázquez, A., 2020. Plant extracts as green corrosion inhibitors for different metal surfaces and corrosive media: A review. *Processes*, *8*(8), p.942.
33. Shang, Z. and Zhu, J., 2021. Overview on plant extracts as green corrosion inhibitors in the oil and gas fields. *Journal of Materials Research and Technology*, *15*, pp.5078–5094.
34. Verma, C., Quraishi, M.A. and Rhee, K.Y., 2021. Present and emerging trends in using pharmaceutically active compounds as aqueous phase corrosion inhibitors. *Journal of Molecular Liquids*, *328*, 115395.
35. El-Shamy, A.M. and Mouneir, S.M., 2023. Medicinal materials as eco-friendly corrosion inhibitors for industrial applications: A review. *Journal of Bio-and Tribo-Corrosion*, *9*(1), p.3.
36. Haque, J. and Quraishi, M.A., 2021. Carbohydrates and their derivatives as corrosion inhibitors. In *Organic Corrosion Inhibitors: Synthesis, Characterization, Mechanism, and Applications* (pp.241–254).Wiley.
37. Eduok, U., 2021. Carbohydrates used as corrosion inhibitors. In *Sustainable Corrosion Inhibitors II: Synthesis, Design, and Practical Applications* (pp. 79–101). American Chemical Society.
38. Khalil, S.M., Ali-Shattle, E.E. and Ali, N.M., 2013. A theoretical study of carbohydrates as corrosion inhibitors of iron. *Zeitschriftfür Naturforschung A*, *68*(8–9), pp.581–586.
39. Shaw, P., Obot, I.B. and Yadav, M., 2019. Functionalized 2-hydrazinobenzothiazole with carbohydrates as a corrosion inhibitor: Electrochemical, XPS, DFT and Monte Carlo simulation studies. *Materials Chemistry Frontiers*, *3*(5), pp.931–940.
40. Rbaa, M., Fardioui, M., Verma, C., Abousalem, A.S., Galai, M., Ebenso, E.E., Guedira, T., Lakhrissi, B., Warad, I. and Zarrouk, A., 2020. 8-Hydroxyquinoline based chitosan derived carbohydrate polymer as biodegradable and sustainable acid corrosion inhibitor for mild steel: Experimental and computational analyses. *International Journal of Biological Macromolecules*, *155*, pp.645–655.
41. Ansari, F.A., Sudheer, Chauhan, D.S. and Quraishi, M.A., 2021. Oleochemicals as Corrosion Inhibitors. In *Organic Corrosion Inhibitors: Synthesis, Characterization, Mechanism, and Applications* (pp.343–369). Wiley.

42. Mohyaldinn, M.E., Lin, W., Gawi, O., Ismail, M.C., Ahmed, Q.A., Ayoub, M.A. and Hasan, A., 2019. Experimental investigation of a new derived oleochemical corrosion inhibitor. *Key Engineering Materials*, *796*, pp.112–120.
43. Saha, S.K., Murmu, M., Murmu, N.C. and Banerjee, P., 2022. Benzothiazolyl hydrazine azomethine derivatives for efficient corrosion inhibition of mild steel in acidic environment: Integrated experimental and density functional theory cum molecular dynamics simulation approach. *Journal of Molecular Liquids*, *364*, 120033.
44. Singh, A., Ansari, K.R., Banerjee, P., Murmu, M., Quraishi, M.A. and Lin, Y., 2021. Corrosion inhibition behavior of piperidinium based ionic liquids on Q235 steel in hydrochloric acid solution: Experimental, density functional theory and molecular dynamics study. *Colloids and Surfaces A: Physicochemical and Engineering Aspects*, *623*, 126708.
45. Ma, I.W., Ammar, S., Kumar, S.S., Ramesh, K. and Ramesh, S., 2022. A concise review on corrosion inhibitors: Types, mechanisms and electrochemical evaluation studies. *Journal of Coatings Technology and Research*, *19*(1), pp.1–28.
46. Aslam, R., Serdaroglu, G., Zehra, S., Verma, D.K., Aslam, J., Guo, L., Verma, C., Ebenso, E.E. and Quraishi, M.A., 2022. Corrosion inhibition of steel using different families of organic compounds: Past and present progress. *Journal of Molecular Liquids*, *348*, 118373.
47. Erteeb, M.A., Ali-Shattle, E.E., Khalil, S.M., Berbash, H.A. and Elshawi, Z.E., 2021. Computational studies (DFT) and PM3 theories on thiophene oligomers as corrosion inhibitors for iron. *American Journal of Chemistry*, *11*, pp.1–7.
48. Rizvi, M., 2021. Natural polymers as corrosion inhibitors. In *Organic Corrosion Inhibitors: Synthesis, Characterization, Mechanism, and Applications* (pp.411–434). Wiley.
49. Zhang, G., Yang, Q., Hou, D., Zhou, P. and Ding, Q., 2022. Unraveling the microstructural properties of cement-slag composite pastes incorporated with smart polymer-based corrosion inhibitors: From experiment to molecular dynamics. *Cement and Concrete Composites*, *125*, 104298.
50. Aribo, S., Olusegun, S.J., Rodrigues, G.L.S., Ogunbadejo, A.S., Igbaroola, B., Alo, A.T., Rocha, W.R., Mohallem, N.D.S. and Olubambi, P.A., 2020. Experimental and theoretical investigation on corrosion inhibition of hexamethylenetetramine [HMT] for mild steel in acidic solution. *Journal of the Taiwan Institute of Chemical Engineers*, *112*, pp.222–231.
51. Ganjoo, R., Sharma, S., Thakur, A., Assad, H., Sharma, P.K., Dagdag, O., Berisha, A., Seydou, M., Ebenso, E.E. and Kumar, A., 2022. Experimental and theoretical study of sodium cocoylglycinate as corrosion inhibitor for mild steel in hydrochloric acid medium. *Journal of Molecular Liquids*, *364*, 119988.

3 Introduction to Sustainable Corrosion Inhibitor

3.1 INTRODUCTION

Metal corrosion is an electrochemical phenomenon arising due to the potential difference in the metal surface. Generally, metals such as mild steel, copper, and aluminum are used for industrial applications, as they are relatively complex and stable and are used as a base material for most construction or equipment. Before these metals are used, rust and scale are removed using various processes such as acid pickling and descaling; HCl, H_2SO_4, and HNO_3 are primarily used for this process [1–4]. Apart from excess rust during these processes, a bare metal surface is also lost, which is especially important to prevent. Typically, corrosion inhibitors are molecules that protect the metal surface from corrosion and maintain their mechanical, electrical, and physical properties. Synthetic organic molecules are predominantly used as excellent potential corrosion inhibitors because they contain electron-donating moieties such as unsaturated pi electrons and heteroatoms (O, S, P, N), because of which they behave like a Lewis base and can donate electrons to the vacant d orbital of the metal surface that acts like a Lewis acid [5–7]. Generally, synthetic green inhibitors are those which can be applied in the environment and are without any adverse effects. These are cost-effective, and primarily, a one-pot multicomponent reaction method is used to make them [8]. Typically, molecules containing hydroxamic acids [9,10], benzothiazole [11,12], Quinoxaline [13,14], imidazole [15,16], benzothiazine [17], chitosan [18,19], etc. are used as synthetic green corrosion inhibitors due to their potential nature and less toxic properties. Currently, theoretical calculation and experimental techniques are used to study the inhibitor molecules' possible properties. Traditional corrosive monitoring techniques encompass gravimetric analysis and electrochemical investigation, which reveal the adsorption properties of any inhibitors and the electrochemical behavior of the inhibitor/metal interface. X-ray diffraction, scanning electron microscopy (SEM), and electron diffraction X-ray spectroscopy are primarily used to investigate morphological changes in the metal surface after applying inhibitor. Apart from all these, recent computer calculation-based theories are applied to explain the molecular structure, electron-donating sites, and adsorption behavior of inhibitors, among which density functional theory (DFT), molecular dynamic (MD), and Monte Carlo (MC) simulation are prominent.

DOI: 10.1201/9781003441151-3

3.2 SUSTAINABLE CORROSION INHIBITORS

Green or sustainable chemistry or process is defined as nonpolluting and safe manufacturing in high yield or preventing pollution through superior synthetic design [20,21]. Greenness is essential for various industrial and household applications, including corrosion inhibition. According to REACH (Registration, Evaluation, Authorization, and Chemical Restriction) and the Oslo Paris Commission (OSPAR), biodegradable and nonbioaccumulative chemicals that have very low or zero environmental toxicity are deemed green or environmentally friendly [22,23]. The toxicity of any substance can be measured using EC50 and LC50 values [24]. EC reflects the effective concentration, and EC50 is the chemical concentration that adversely affects the growth of 50% of a living population.

Conversely, LC reflects the lethal concentration, and LC50 is the chemical concentration that would kill 50% of the living population. It is estimated that for a chemical to be green, EC50/LC50 value should be more excellent than 10 mg L21. According to OSPAR, an acceptable biodegradation level for any chemical is 60% in 28 days [25,26]. Bioaccumulation is deliberate in partition coefficient (log KOW or DOW), a gauge of chemical distribution in a water and octanol mixture. For a chemical to be green, log KOW should be less than 3. It is widely known that there is no proper definition associated with the term "environment friendly" or "green" corrosion inhibitors. Because of their cost-effectiveness and versatility, inorganic inhibitors (mostly chromates) were used alone as the earliest, most versatile, and most effective corrosion inhibitors for ferrous and nonferrous alloys. Primarily, inorganic inhibitors function as an anodic or passivator type of inhibitors. A critical concentration is highly required for better effectiveness, as chromate causes pitting corrosion below the critical concentration. Inorganic inhibitors are recognized as the "gold standard" in corrosion inhibition; however, recently, their use has been phased out because of their toxicity and expensive discharge [27, 28]. The increasing ecological sensitivity and awareness restrict the consumption of traditional toxic inorganic and organic corrosion inhibitors. Recently, environmentally sustainable corrosion inhibitors from nontoxic and biodegradable sources are becoming more in demand compared to commercially available toxic organic and inorganic inhibitors [29].

To select the compounds to perform as corrosion inhibitors, several other factors are also to be considered, such as their availability and economic viability. Given these factors, the current research in developing the corrosion inhibitor is directed toward evaluating nontoxic, environmentally sound, and cost. Green-based corrosion inhibitors are further divided into organic and inorganic classes. Organic inhibitors are nontoxic; such as flavonoids, phenols, alkaloids, and by-products of waste plants, while inorganic inhibitors are toxic up to a specific concentration and thus need to be functionalized before their usage. Natural and green corrosion inhibitors are plant extracts, polymers, natural gums, drugs, plant wastes, etc. Recently, synthetic corrosion inhibitors have been constrained globally due to the increasing awareness of harsh environmental regulations and public health. For this reason, the attention of scientists and chemical engineers was attracted toward the synthetic development of such corrosion inhibitors, which are environment friendly, less toxic, inexpensive, and do not involve a complicated process when making and using them [30, 31]. The increasing number of publications on corrosion inhibition of various metals during the past decade suggests the interest in exploring new inhibitors for different corrosive environments. However, the number of publications on green corrosion inhibitors is limited. So far, several organic and inorganic compounds have been evaluated as green and eco-friendly inhibitors, and much work has been devoted to developing efficient green corrosion inhibitors extracted from natural plants [32–35].

Plant extracts and drug molecules are environmentally benign and ecologically acceptable compounds, economically viable, readily available, and biodegradable. Not only does the leaf portion of the plant inhibit corrosion, but the root, stem, flower, seed, fruit, bark, and peel may also act as excellent corrosion inhibitors [36–38]. However, leaves have been preferred most because of their abundant phytochemicals and other active components. These inhibitors are used in relatively low concentrations to inhibit metal surface corrosion in aggressive environments. Plant extracts and drug molecules are commonly used extensively as green corrosion inhibitors because their molecules have electron-donating abilities and sites present. They are safe to use, biodegradable, sustainable, and can be easily obtained. Their inhibition efficiency is attributable to the complex organic substances involved in their formulation (tannins, alkaloids, nitrogen bases, carbohydrates, proteins, and some hydrolysis products). They typically include heteroatoms in their molecular structures, such as nitrogen, sulfur, or oxygen atoms, as well as triple or conjugated double bonds or aromatic rings. The ionic liquid is in this category, which behaves like a green potential corrosion inhibitor [39]. Most synthetic organic molecules involve very complex multiple approaches to use. They threaten the environment, but some synthetic organic molecules are less toxic, water-soluble, eco-friendly, and cheap. Depositing them on a metal surface protects the metal from further corrosion in corrosive media. Heterocyclic compounds and polymers commonly come in the form of synthetic green corrosion inhibitors, of which indole [40], triazole [41], benzimidazole [42], thiols [43], imidazole [44], pyrimidine [45], etc. and their derivative are of prime interest. Hence, they have been widely used for the last few years as green and sustainable corrosion inhibitors for metal and material protection.

Polymers, both natural and synthetic, also have been studied in various corrosive conditions as corrosion inhibitors for different materials and therefore have gained significant popularity. These polymers work by blanketing the metal surface via the heteroatoms present in their structure. They are chemically stable, inexpensive, and renewable sources [46]. Similar to plant extracts and drug molecules, they have also been found to be attractive as corrosion inhibitors as they are cost-effective, inherently stable, and non-toxic. Similarly, natural and synthetic polymers have received wide acceptance and have been assessed as corrosion inhibitors in different corrosion environments for various metals. The applications of polymer are also cost-effective, stable, and non-toxic. Compared to monomer analogs, polymers demonstrate a superior inhibition effect because they possess long-chain carbon linkage and multiple adsorption sites and thus block the large surface area of the corroding metal, thereby protecting the metal from corrosive agents present in the solution. Polymers have two significant advantages:

- A single polymer chain displaces several water molecules from the metallic surface, thus making the process entropically significant.
- Multiple bonding sites make the desorption of polymer molecules a slower process.

The protection offered by the polymers was observed to be related to their chemical composition, molecular weight, and electronic structures. The polymeric inhibitors, which have been investigated as corrosion inhibitors include polyethylene glycol, polyvinyl alcohol, polyvinyl pyrrolidone, polyethyleneimine, polyacrylic acid, polyacrylamide, starch, gum arabic, guar gum, xanthan gum, gellan gum, hydroxypropyl cellulose, hydroxyethyl cellulose, psyllium polysaccharide, carboxymethyl cellulose, pectin, and chitosan.

Consequently, corrosion scientists and engineers are frequently paying attention to the expanding eco-friendly and sustainable alternatives of either artificial or natural origins. In this course, one-step multicomponent reactions (MCRs), during which three or more reactants are combined in a single step, have been deemed as one of the greenest protocols for developing environmentally benign alternative corrosion inhibitors. MCRs are interrelated using many profitable features, including high synthetic yield, ease of operation, and lower number of purification and workup steps, particularly in catalyst and solvent-free situations. One study suggested that numerous green inhibitors derived using MCRs are successfully employed as effective corrosion inhibitors for metals and alloys. A detailed description of compounds synthesized using MCRs as green corrosion inhibitors is published elsewhere [47]. Nowadays, chemicals synthesized using microwave (MW) and ultrasound (US) irradiations, especially in association with MCRs, are considered as another green protocol as these nonconventional heating techniques are allied using plentiful advantageous uniqueness together with instantaneous heating of reaction mixture, generation of high temperature, and high chemical selectivity. Owing to the association of MW and US irradiations with environmental friendliness, several heterocyclic compounds are synthesized and evaluated as effective corrosion inhibitors. Chemical syntheses using biomass and biologically originated chemicals such as amino acids and carbohydrates and their derivatives can also be considered as green corrosion inhibitors. Apart from these, water and supercritical carbon dioxide can be considered environment-friendly solvents; therefore, chemicals synthesized using the solvents are regarded as green chemicals and can be used as nontoxic alternatives for traditional corrosion inhibitors. The high thermal and chemical stability, high solubility in polar medium, low vapor pressure, and low volatility make the ionic liquids (ILs) a suitable alternative source of corrosion inhibitors.

3.2.1 Adsorption Mechanism of Sustainable Corrosion Inhibitors

The general assumption is that organic molecules' first step in the corrosion inhibition mechanism of metals is their adsorption onto the metal surface via their functional group. Therefore, to understand the underlying corrosion inhibition mechanisms, inhibitors' adsorption behavior on the metal surface must be known [48, 49]. The adsorption process depends upon the following factors:

- The topography, uniformity, and physicochemical properties (charge, conformation, chemistry) of the metal surface
- The propagation of the charge in the molecule
- The aggressive medium, and
- The chemical composition of the organic molecule.

The adsorption of ions or neutral molecules on bare metal surfaces immersed in solution is determined by the mutual interactions of all species present at the phase boundary. These include electrostatic and chemical interactions of the adsorbate with the surface, adsorbate–adsorbent and adsorbate–solvent interactions. Two modes of adsorption are present depending on the existence of the forces involved: Physisorption is where a corrosive anion, such as Cl, SO_4, PO_4, or NO, forms a layer above the metal surface in the presence of a highly acidic medium and a protonated lone pair containing heteroatom present in the inhibitor molecule. Here there is an interaction between the protonated inhibitor molecules and the deposit of the anion above the metal surface. Earlier studies suggested that if the value of Gibbs adsorption energy (ΔGads) is more than −20 kJ/mol, it is usually kept under physisorption [50].Conversely, chemisorption is a phenomenon without direct interaction between inhibitor molecules and the metal surface. The unshared and unsaturated pi-electrons in the inhibitor molecules form a strong base by interacting with the metal surface's vacant "d" orbital. They are deposited above the metal surface as a thin protective layer. Previous studies suggested that if the value of Gibbs adsorption energy (ΔGads) is more than −40 kJ/mol, it is usually kept under chemisorptions [50]. If seen, corrosion inhibition is a kind of acid-base-based mechanism in which the inhibitor behaves like a Lewis base; that is, it can donate its electron, whereas a metal Lewis acts like acid, which accepts an electron.

- Chemisorptions: In this type of adsorption, a single layer of molecules, atoms, or ions is attached to the surface by chemical bonds and is essentially irreversible; chemisorption has a higher heat of adsorption than adsorption involving electrostatic/van der Waals forces. Chemisorption commonly takes place through the donation of electrons from species with loosely bonded electrons, such as π-electrons in aromatic rings, or multiple bonds, or unpaired electrons in functional groups that contain atoms such as oxygen, nitrogen, sulfur, and phosphorus, to the vacant d-orbital of transition metals.
- Physical adsorption: In this adsorption mode, the bonding is by the weaker van der Waals forces, whose energy levels are comparable to condensation. In the case of physisorption, coulombic interactions between ions or dipoles of the inhibitor and charged metal surface are present, and other interactions also exist (van der Waals forces, hydrogen bonding). In this case, the adsorption process is rapid, but the adsorbed species can easily be removed from the surface.

The performance of corrosion inhibition of the sequence homology set of organic substances differing only in the heteroatom is usually as follows: phosphorus, sulfur, nitrogen, and oxygen. Factors influencing the absorption of ion inhibitors on metal surfaces are as follows:

- Metal surface charge: Adsorption may arise from the electrostatic force of attraction between the ions or dipole loads on the adsorbed inhibitor and the electrical charge on metal at the interface of the metal solution.
- The functional groups and structure of inhibitor: By electron transfer, inhibitors can bind to the metal surface and form a coordinated bond that contributes to good binding and successful inhibition. Species with comparatively loosely bound electrons in anions, neutral compounds, lone pairs of electrons, triple bond or organic ring systems-related π-electron systems, and functional groups with periodic table elements of group V or VI prefer fast transfer of electrons and better bond formation and thus successful inhibition. The propensity to form stronger coordinate bonds increases with declining electronegativity and follows the order of oxygen, nitrogen, sulfur, and phosphorus.
- The interaction between adsorbed inhibitor species (synergism and antagonism): Adsorbed species may enter into various interactions on the surface of an electrode that may significantly influence their inhibitive properties and the mechanism of their action.
- The reaction of adsorbed inhibitors: The adsorbed inhibitor species may usually react by the electrochemical reaction to form an inhibitive product. Inhibition due to the added substance is termed primary inhibition, and secondary inhibition is due to reaction products.

The adsorption to metal surfaces of organic inhibitors is widely known as isothermal adsorption. Adsorption isotherms are widely acknowledged to offer invaluable insights into the mechanism of corrosion inhibition and provide essential details about the nature of interactions amongst the metals and inhibitors. It is possible to deduce the strength of the adsorption bond from the adsorption isotherm, which shows the contrast between the inhibitor surface concentration and bulk solution. To illustrate the mode of adsorption of inhibitors onto the metal surface, various isotherms of adsorption such as Langmuir, Temkin, Frumkin, Flory–Huggins, Dhar–Flory–Huggins, Bockris–Swinkels, Freundlili–Swinkels, and Freundlich adsorption isotherms have been reported.

3.2.2 Factors Influencing the Efficiency of Corrosion Inhibitors

Several factors should be considered to properly select the inhibitors, among which the primary decision is to match the appropriate inhibitor chemistry with the corrosion conditions (Figure 3.1). The selection of the inhibitors starts with the choice of the physical properties, that is, whether the inhibitor should be in a solid or liquid state. The importance of the melting and freezing points must also be considered. Another issue that must also be considered is whether their degradation with time and temperature

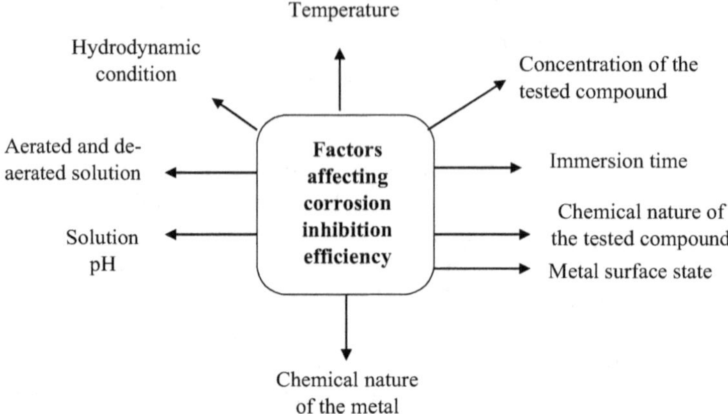

FIGURE 3.1 Factors which determine the performance of the corrosion inhibitors [51].

is critical, which should be compatible with other system additives and specific solubility characteristics as required. This list can be extensive, but it is crucial because it paves the way for possible corrosion inhibitors. For the evaluation of any new system, it is considered as the first step. The physical quantifications are those customarily done as part of the least quality affirmation testing. The simplest corrosion test should be executed to determine the inhibitors' performance and rule out unsuitable candidates. The aim of the initial screening is that the inhibitors which perform poorly must not be carried out further. The inhibitor user must employ test procedures that rigorously exclude inferior inhibitors, although some suitable inhibitors may also be excluded. The most severe challenge in evaluating the performance of the inhibitor system is designing the conditions (real world) of the experiments needed to estimate their performance. The variables that must be considered include temperature, pressure, velocity, metal properties, and corrosive environment chemistry.

Corrosion inhibition requires the inhibitive species to easily access the metal surface. Therefore, surfaces should be clean and not contaminated by oil, grease, corrosion products, water hardness scales, etc. Further measures should be taken to avoid the deposition of solid particles. This conditioning is often challenging to achieve, and there are many cases where less than adequate consideration has been given to the preparation of systems to receive inhibitive treatment. It is also necessary to ensure the inhibitor reaches all parts of the metal surfaces. Care should be taken, particularly when first filling a system, that all the inhibited fluid regions contact the regions. The movement of the fluid in service will encourage this in many systems. Still, in nominally static systems, it will be desirable to establish a flow regime at intervals to provide a renewed supply of inhibitors. Inhibitors must be selected after considering the nature and combinations of metals present, the nature of the corrosive environment, and the operating conditions in terms of flow, temperature, and heat transfer. Inhibitor concentrations should be checked regularly, and losses restored either by appropriate additions of inhibitor or by complete replacement of the whole fluid. Some form of continuous monitoring should be employed. However, it must be remembered that the results from monitoring devices, probes, coupons, and so forth refer to the behavior of that particular component at that specific part of the system. Nevertheless, despite this caution, it must be recognized that corrosion monitoring in an inhibited system is well-established and widely used.

3.3 INDUSTRIAL APPLICATIONS OF CORROSION INHIBITORS

There is a range of other uses of the applicability of corrosion inhibitors such as acid pickling, gasoline, gas, and petroleum industries, water recirculation systems, chemical processing, water treatment, and additive materials industries [52]. In the oil production, refining, and chemical fields, the first line of protection is applying corrosion inhibitors. The other application of corrosion inhibitors is in the antifreeze of automobiles [53]. Several applications of corrosion inhibitors are depicted in Figure 3.2.

3.3.1 WATER

Potable water is corrosive unless a protective film or deposit is formed on the surface of metal or alloys. Potable water is usually saturated with dissolved oxygen. Some of the commonly utilized corrosion inhibitors for potable water are:

- $Ca(HCO_3)_2$: used to protect cast iron and steel (concentration of 10 ppm).
- Polyphosphate: utilized to protect Fe, Zn, Cu, and Al (concentration of 5–10 ppm).
- $Ca(OH)_2$: utilized to protect Fe, Zn, and Cu (concentration of 10 ppm).
- Na_2SiO_3: utilized to protect Fe, Zn, and Cu (concentration of 10–20 ppm).

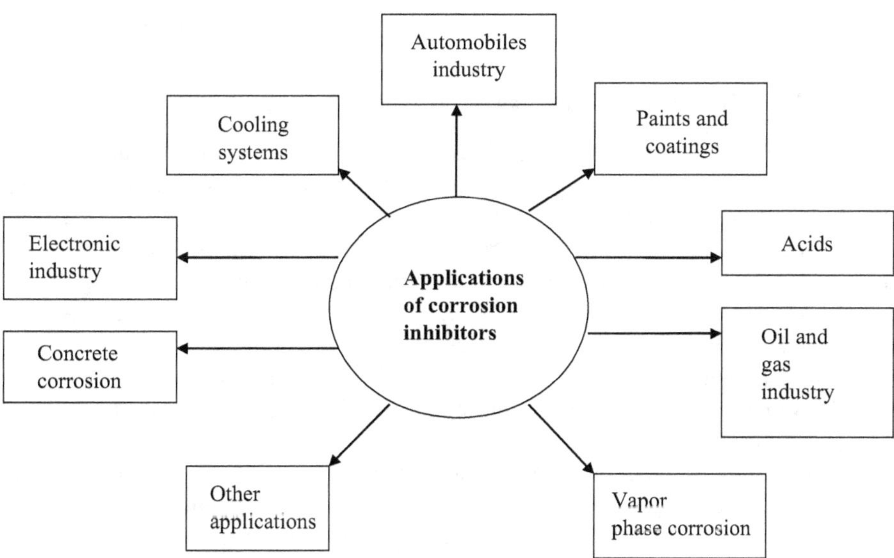

FIGURE 3.2 Example of the applications of corrosion inhibitors.

3.3.2 Cooling Waters

In a recirculating system, evaporation is the primary source of cooling. As evaporation proceeds, the dissolved mineral salt content increases. Cooling systems may comprise several dissimilar metals and non-metals. Metals picked up from one part of the system can be deposited elsewhere, producing galvanic corrosion. Corrosion is controlled by anodic (passivating) inhibitors, including nitrate and chromate. Cathodic (e.g., zinc salt) inhibitors and organic corrosion inhibitors are also used. Some of the most utilized corrosion inhibitors in the cooling water system are listed below:

- $Ca(HCO_3)_2$: used to protect cast iron and steel (concentration 10 ppm).
- Na_2CrO_4: utilized to protect Fe, Zn, and Cu (concentration 0.1%).
- $NaNO_2$: utilized for the protection of Fe (concentration 0.05%).
- NaH_2PO_4: utilized to protect Fe (concentration 1%).
- Morpholine: utilized for the protection of Fe (concentration 0.2%).

3.3.3 Boiler Water

Inhibitors are used to control corrosion in boiler waters; some of the commonly used inhibitors are:

- NaH_2PO: utilized to protect Fe, Zn, and Cu (concentration 10 ppm).
- Polyphosphate: utilized for the protection of Fe, Zn, and Cu (concentration 10 ppm).
- Morpholine: used to protect Fe (used at variable concentration).
- Hydrazine: used to protect Fe (used as an O_2 scavenger).
- Ammonia: utilized for the protection of Fe (used as neutralizer).
- Octadecylamine: used to protect Fe (used at variable concentrations).

3.3.4 Acid Pickling and Boilers

Some of the significant applications of acid, in which metals or their alloys are in direct or indirect contact, are acid picking, acid well acidizing processes, cleansing of oil gas industrial equipment, and heat exchangers. Mixed types of corrosion inhibitors are employed primarily to protect them from corrosion. Some of the most used inhibitors in an acidic environment are summarized as follows:

- Ethylene: protects Fe in HCl medium (0.5% concentration).
- Mercaptobenzotriazole: protects Fe in HCl medium (1% concentration).
- Pyridine phenylhydrazine: protects Fe in HCl medium (0.5–10.5% concentration).
- Rosin amine-ethylene oxide: used to protect Fe in HCl medium (0.2% concentration).
- Phenylacridine: protects Fe in H_2SO_4 medium (0.5% concentration).
- NaI: protects Fe in H_3PO_4 medium (200 ppm concentration)

3.3.5 Automobile

Inhibitors are used to minimize internal corrosion, usually caused by coolant aeration temperature, pressure, etc. Nitrites, nitrates, phosphates, silicates, arsenates, chromates (anodic inhibitors), amines, benzoates, mercaptans, organic phosphates (mixed inhibitors), and polar or emulsifiable oils (film formers) are some of the widely used antifreeze corrosion inhibitors. Another use of corrosion inhibitors in automobiles is to protect against corrosion of metal surfaces exposed to the environment. The atmosphere to which automobiles are exposed contains moist air, wet SO_2 gas (forming sulfuric acid in the presence of moist air), and deicing salt (NaCl and $CaCl_2$). The corrosion-proofing formulations used include grease, wax resin, resin-emulsion, and metallo-organic and asphaltic compounds to control external corrosion. Typical inhibitors used in rust-proofing applications are fatty acids, phosphonates, sulfonates, and carboxylates.

3.3.6 Organic Coatings and Paints

In the primers, finely powdered inhibiting pigments are incorporated. They usually displace the water and orient themselves, so their hydrophobic tails are towards the environment. Red lead (Pb_3O_4) is commonly used in paints on iron. It helps to preserve the physical properties of the paints. Lead azelate, calcium plumbate, and lead suboxide are other utilized inhibitors.

3.3.7 Internal Corrosion of Steel Pipelines

The assembly pipelines between the oil and gas wells and the processing plants face the same problem as the refineries and petrochemical plants. The flow of the multiphase fluids affects the corrosion rate of the pipelines. At exceedingly high rates, flow-induced corrosion and erosion-corrosion occur; the increased flow tends to cause the sweep of the sediments out of the pipelines. However, at low flow rates, the usual form is pitting, as low velocity is related to settling the deposits at the bottom. The corrosion of the pipelines or pigging is regulated by cleaning them by adding continuous or batches of corrosion inhibitors.

3.3.8 Petroleum Production

Refineries and petrochemical industries employ film-forming inhibitors to manage wet corrosion. Several inhibitors, including amines and amides, are long-chain nitrogenous organic materials. Water-soluble and water-soluble dispersible oil inhibitors and oil-soluble and oil-soluble water dispersible inhibitors (batch inhibitors) are continuously incorporated to mitigate corrosion problems. They are active at liquid–liquid and/or liquid–gas interfaces as inhibitors are interfacial and can lead to emulsification. Consequently, in the presence of inhibitors, foaming is rarely observed. Film-forming inhibitors; anchor to the metal via their polar group, and the nonpolar tail protrudes out vertically. The physical adsorption of hydrocarbons (oils) on such nonpolar tails improves film thickness and the efficacy of hydrophobic barriers for corrosion inhibition.

3.3.9 Corrosion Inhibition in Oil and Petroleum Industries

In the early days, many chemical compounds were added by oil companies, often effectively, to mitigate corrosion damage to oil wells and surface handling machinery. Corrosion prevention in oil fields became significantly more effective after using amines and imidazolines. Current inhibitors are introduced in the area at concentrations of 15–50 ppm, constantly or in intermittent batches, based on overall liquid output. A much broader range of inhibitor chemistry is available today than decades ago to tackle oil-field corrosion. In recent years, organic molecules containing sulfur, phosphorus, and nitrogen in various combinations have been developed. These types of inhibitors have increased the efficiency of oil field inhibitors and control corrosion associated with high CO_2 and low H_2S conditions. Most of the inhibitors currently used in developing wells are organic nitrogenous compounds. As part of the structure, the primary forms have long-chain hydrocarbons (C_{18}). Many of the inhibitors being used effectively today are either with long-chain aliphatic diamine or imidazolines from the long carbon chain. Numerous modifications of these structures are being introduced to alter the material's physical properties, such as the typical reaction of ethylene oxide with these compounds in different molecular percentages to provide polyoxyethylene derivatives with varying degrees of miscibility of the salt. A much broader range of inhibitor chemistry is available today to tackle oilfield corrosion; water or water salt solutions alone will not cause harmful corrosion unless unique corrodents are contained, such as CO_2, H_2S, and their dissolution products. The wells for oil and gas are either sweet or sour [54]:

- Sweet corrosion: Corrosion can be grouped into three ambient temperatures in CO_2 gas wells. The corrosion product is nonprotective under 60°C, and there will be high corrosion rates. Magnetite is formed above about 150°C, and the wells are not corrosive even with high brine levels. The iron carbonate corrosion product layer is protective in the middle-temperature region, where most gas well conditions lie, but is adversely affected by chlorides and fluid velocity. The dispersibility feature of the oil and brine being produced are among the oilfield inhibitors' significant physical properties. A properly chosen inhibitor based on the corrosion mechanism will only be successful if it has access to the corrosive metal. There are also some significant variations in handling oil and gas wells.

  There is a need for a clear cut in the difference between an oil well and a gas well. The fact that a large volume of gas is generated by many oil wells and a large volume of liquid is produced by many gas wells and that wells frequently undergo a shift in production over their lifetime makes it challenging to create a technical distinction. Typical gas wells are much hotter than oil wells and are lighter regarding hydrocarbon liquids. Typically, gas wells are more profound and produce lower total dissolved solids brines. Oxygen is not a consideration in gas well corrosion but may cause significant issues in artificial lift oil wells. Corrosion mechanisms may change due to the wide temperature gradient in many gas wells, resulting in various forms of corrosion in the same well. Conversely, oil wells do not demonstrate this behavior. Oil wells usually contain more liquid than gas wells, resulting in a shorter batch treatment. As corrosion in oil wells is electrochemical, an electrolyte must be present for corrosion to occur. The water source in oil wells is usually the formation source, and at concentrations ranging from trace to saturation, the water may contain dissolved salts. Corrosion-related water can be in a thin layer, droplets, or a significant process. The corrosion control analysis results by inhibitors in developing oil wells in fields flooded with carbon dioxide showed that imidazolines effectively protect CO_2 brines. Inhibitors, such as nitrogen-phosphorus or sulfur compounds in organic molecules, also performed better.

- Sour corrosion: Aldehydes, cyanamide thiourea, and urea derivatives are corrosion inhibitors used in the past to tackle the problem of corrosion faced by sour wells. Organic amines are the most frequently employed inhibitors in sour wells. While organic amines are assumed to be somewhat less effective inhibitors of the acid solution, amine inhibition is significantly accelerated by the presence of hydrogen sulfide [55]. Hydrogen sulfide is the principal corrosive agent in sour wells, and carbon dioxide is also often present. The effect of multiple iron sulfides at varying hydrogen sulfide concentrations in corrosion products has also been established. Oilfield inhibitors integrate corrosion products on the metal surface into a thin layer. This surface film can be anaerobic or partly oxidized and may be sulfide or carbonate.
- Acidizing: acidizing is a practical approach for promoting the production of oil and gas wells. These cannot flow readily into the well due to the very poor permeability of such hydrocarbon-containing formations. If the rock is sandstone, these limestone/dolomite formations can be treated with HCl or an HF mixture. In the acidifying procedure, the acid is pumped down the tubing into the well, reaching the performance and hitting the structure; the acid etches channels that provide a path for oil and gas to enter the pipe. Several inhibitors, such as high-molecular-weight nitrogenous compounds, alkyl or alkylaryl nitrogen compounds, and acetylenic alcohols, such as 1-octyn-3-ol, are utilized for well-acidizing operations. However, as they are very hazardous, these materials pose severe handling problems; this will determine which product is ultimately used by an operator. In addition, both in productivity and time, their effectiveness is limited. Acid soaks usually last between 12 and 24 hours, after which time the efficiency of the inhibitor can start to drop alarmingly. Cinnamaldehyde and alkynols containing unsaturated groups in conjunction with the oxygen role defined as alpha-alkenylphenones are oxygen-containing inhibitors that are efficient in concentrated HCl. They provide protection similar to that obtained from acetylenic alcohols, especially when combined with small amounts of surfactants.

3.3.10 Others

Inhibitors have been used for corrosion mitigation in fuel oil reservoirs, hot chloride dye baths, cooling brines, and steel reinforcement in concrete. Several corrosion inhibitors are also used to protect artifacts from corrosion.

3.4 SUMMARY

As discussed above, using corrosion inhibitors is one of the most prevalent ways of combating corrosion in various industries. For decades, conventional corrosion inhibitors have been in practice; however, due to their toxic behavior toward the environment, their usage has remained restricted. Thus, using environment-friendly natural inhibitors is in practice as they are effective, readily available, and economically feasible. The above discussion delineates that green corrosion inhibitors are predominantly applied toward corrosion mitigation of metals and alloys in various corrosive solutions. They are used primarily because of their potential behavior, water solubility, and corrosive media stability at different temperatures. They work primarily by getting adsorbed onto the surface of the metals, forming a thin barrier onto the metal surface. One of the challenges associated with corrosion inhibitors is their selection. Proper choice of inhibitors should be made by matching the appropriate inhibitor chemistry with the corrosion conditions and selecting suitable physical properties for the application conditions. Method of application and system characteristics must be taken into account when determining the physical properties of an inhibitor. Inhibitors must be selected considering the metals present, the nature of the corrosive environment, and the operating conditions in terms of flow, temperature, and heat transfer. Inhibitor concentrations should be monitored regularly, and losses restored either by appropriate additions of inhibitor or by complete replacement of the whole fluid as recommended.

REFERENCES

1. Verma, C., Quraishi, M.A. and Ebenso, E.E., 2020. Corrosive electrolytes. *International Journal of Corrosion and Scale Inhibition*, *9*(4), pp.1261–1276.
2. Shinato, K.W., Zewde, A.A. and Jin, Y., 2020. Corrosion protection of copper and copper alloys in different corrosive medium using environmentally friendly corrosion inhibitors. *Corrosion Reviews*, *38*(2), pp.101–109.
3. Baitule, P. and Manivannan, R., 2021. Corrosion protection in acidic medium for mild steel using various plants extract as green corrosion inhibitor-a review. *Key Engineering Materials*,*882*, pp.50–63.
4. Zhang, Q., Zhang, R., Wu, R., Luo, Y., Guo, L. and He, Z., 2022. Green and high-efficiency corrosion inhibitors for metals: a review. *Journal of Adhesion Science and Technology*, pp.1–24.
5. Wei, X., Luo, X., Wu, N., Gu, W., Lin, Y. and Zhu, C., 2021. Recent advances in synergistically enhanced single-atomic site catalysts for boosted oxygen reduction reaction. *Nano Energy*,*84*, 105817.
6. Basik, M. and Mobin, M., 2020. Chondroitin sulfate as potent green corrosion inhibitor for mild steel in 1 M HCl. *Journal of Molecular Structure*,*1214*, 128231.

7. Adam, M.S.S., Soliman, K.A. and Abd El-Lateef, H.M., 2020. Homo-dinuclear VO_2^+ and Ni_2^+ dihydrazone complexes: Synthesis, characterization, catalytic activity and CO_2-corrosion inhibition under sustainable conditions. *Inorganica Chimica Acta, 499*, 119212.
8. Machado, I.V., Dos Santos, J.R., Januario, M.A. and Corrêa, A.G., 2021. Greener organic synthetic methods: Sonochemistry and heterogeneous catalysis promoted multicomponent reactions. *Ultrasonics Sonochemistry, 78*, 105704.
9. Wang, H., Hu, L., Cao, G., Xia, R., Cao, J., Zhang, J. and Pan, G., 2022. Experimental and computational studies on octyl hydroxamic acid as an environmentally friendly inhibitor of cobalt chemical mechanical polishing. *ACS Applied Materials & Interfaces, 14*(24), pp.28321–28336.
10. Xia, R., Hu, L., Cao, J., Pan, G. and Qi, Y., 2023. Characterization of octyl hydroxamic acid as inhibitor on Cu chemical mechanical polishing. *ECS Journal of Solid State Science and Technology, 12*(5), 054003.
11. Farag, A.A., Eid, A.M., Shaban, M.M., Mohamed, E.A. and Raju, G., 2021. Integrated modeling, surface, electrochemical, and biocidal investigations of novel benzothiazoles as corrosion inhibitors for shale formation well stimulation. *Journal of Molecular Liquids, 336*, 116315.
12. Liu, Z., Fan, B., Zhao, J., Yang, B. and Zheng, X., 2023. Benzothiazole derivatives-based supramolecular assemblies as efficient corrosion inhibitors for copper in artificial seawater: Formation, interfacial release and protective mechanisms. *Corrosion Science*, 110957.
13. Quadri, T.W., Olasunkanmi, L.O., Fayemi, O.E., Lgaz, H., Dagdag, O., Sherif, E.S.M., Alrashdi, A.A., Akpan, E.D., Lee, H.S. and Ebenso, E.E., 2022. Computational insights into quinoxaline-based corrosion inhibitors of steel in HCl: Quantum chemical analysis and QSPR-ANN studies. *Arabian Journal of Chemistry, 15*(7), 103870.
14. Laabaissi, T., Benhiba, F., Missioui, M., Rouifi, Z., Rbaa, M., Oudda, H., Ramli, Y., Guenbour, A., Warad, I. and Zarrouk, A., 2020. Coupling of chemical, electrochemical and theoretical approach to study the corrosion inhibition of mild steel by new quinoxaline compounds in 1 M HCl. *Heliyon, 6*(5), e03939.
15. Ouakki, M., Galai, M. and Cherkaoui, M., 2022. Imidazole derivatives as efficient and potential class of corrosion inhibitors for metals and alloys in aqueous electrolytes: a review. *Journal of Molecular Liquids, 345*, 117815.
16. Zeng, X., Zheng, X., Guo, L., Xu, Q., Huang, H. and Tan, B., 2021. Three imidazole ionic liquids as green and eco-friendly corrosion inhibitors for mild steel in sulfuric acid medium. *Journal of Molecular Liquids, 324*, 115063.
17. Benhiba, F., Sebbar, N.K., Bourazmi, H., Belghiti, M.E., Hsissou, R., Hökelek, T., Bellaouchou, A., Guenbour, A., Warad, I., Oudda, H. and Zarrouk, A., 2021. Corrosion inhibition performance of 4-(prop-2-ynyl)-[1, 4]-benzothiazin-3-one against mild steel in 1 M HCl solution: experimental and theoretical studies. *International Journal of Hydrogen Energy, 46*(51), pp.25800–25818.
18. Verma, C., Quraishi, M.A., Alfantazi, A. and Rhee, K.Y., 2021. Corrosion inhibition potential of chitosan based Schiff bases: Design, performance and applications. *International Journal of Biological Macromolecules, 184*, pp.135–143.
19. Zhao, Q., Guo, J., Cui, G., Han, T. and Wu, Y., 2020. Chitosan derivatives as green corrosion inhibitors for P110 steel in a carbon dioxide environment. *Colloids and Surfaces B: Biointerfaces, 194*, 111150.
20. Farooq, S., Wu, H., Nie, J., Ahmad, S., Muhammad, I., Zeeshan, M., Khan, R. and Asim, M., 2022. Application, advancement and green aspects of magnetic

molecularly imprinted polymers in pesticide residue detection. *Science of the Total Environment,804*, 150293.

21. Das, P.P., Muduli, M., Borah, S. and Chaudhary, V., 2021. Principle of green chemistry: A modern perspective for development of sustainable textile fiber-based green nanocomposites. In *Green Chemistry for Sustainable Textiles* (pp.121–136). Woodhead Publishing.
22. Verma, C. and Quraishi, M.A., 2021. Green corrosion inhibitors derived from synthesis: Progress and future directions. In *Sustainable Corrosion Inhibitors II: Synthesis, Design, and Practical Applications* (pp.121–147). American Chemical Society.
23. Sosa, Y., 2022. Acceleration to lubricants sustainability. *Tribology & Lubrication Technology*, *78*(2), pp.44–50.
24. Ott, A., Martin, T.J., Acharya, K., Lyon, D.Y., Robinson, N., Rowles, B., Snape, J.R., Still, I., Whale, G.F., Albright III, V.C. and Bäverbäck, P., 2020. Multi-laboratory validation of a new marine biodegradation screening test for chemical persistence assessment. *Environmental Science & Technology*, *54*(7), pp.4210–4220.
25. Fritt-Rasmussen, J., Møller, E.F., Kyhn, L.A., Wegeberg, S., Lassen, P., Cooper, D. and Gustavson, K., 2021. Biodegradation, bioaccumulation and toxicity of oil spill herding agents in Arctic waters as part of an ecotoxicological screening. *Water, Air, & Soil Pollution*, *232*, pp.1–14.
26. Fritt-Rasmussen, J., Møller, E.F., Kyhn, L.A., Wegeberg, S., Lassen, P., Cooper, D. and Gustavson, K., 2021. Biodegradation, bioaccumulation and toxicity of oil spill herding agents in Arctic waters as part of an ecotoxicological screening. *Water, Air, & Soil Pollution*, *232*, pp.1–14.
27. Verma, C. and Quraishi, M.A., 2021. Green corrosion inhibitors derived from synthesis: Progress and future directions. In *Sustainable Corrosion Inhibitors II: Synthesis, Design, and Practical Applications* (pp.121–147). American Chemical Society.
28. Husna, U.Z., Elraies, K.A., Shuhili, J.A.B. and Elryes, A.A., 2022. A review: the utilization potency of biopolymer as an eco-friendly scale inhibitors. *Journal of Petroleum Exploration and Production Technology*, pp.1–20.
29. Veedu, K.K., Banyangala, M., Kalarikkal, T.P., Somappa, S.B. and Gopalan, N.K., 2023. Green approach in anticorrosive coating for steel protection by *Gliricidia sepium* leaf extract and silica hybrid. *Journal of Molecular Liquids*, *369*, 120967.
30. Fetouh, H.A., Hefnawy, A., Attia, A.M. and Ali, E., 2020. Facile and low-cost green synthesis of eco-friendly chitosan-silver nanocomposite as novel and promising corrosion inhibitor for mild steel in chilled water circuits. *Journal of Molecular Liquids*, *319*, 114355.
31. El Ibrahimi, B., Jmiai, A., Bazzi, L. and El Issami, S., 2020. Amino acids and their derivatives as corrosion inhibitors for metals and alloys. *Arabian Journal of Chemistry*, *13*(1), pp.740–771.
32. Bilgiç, S., 2021. Plant extracts as corrosion inhibitors for mild steel in HCL media-review I. *International Journal of Corrosion and Scale Inhibition*, *10*(1), pp.145–175.
33. Miralrio, A. and Espinoza Vázquez, A., 2020. Plant extracts as green corrosion inhibitors for different metal surfaces and corrosive media: a review. *Processes*, *8*(8), p.942.
34. Wang, X., Chen, L., Yang, F., Xiang, Q. and Liu, J., 2023. Corrosion inhibition mechanism and extraction technology of plant corrosion inhibitors: a review. *Journal of Adhesion Science and Technology*, *37*(21), pp.1–25.

35. Zakeri, A., Bahmani, E. and Aghdam, A.S.R., 2022. Plant extracts as sustainable and green corrosion inhibitors for protection of ferrous metals in corrosive media: A mini review. *Corrosion Communications*, *5*, pp.25–38.
36. Chaubey, N., Qurashi, A., Chauhan, D.S. and Quraishi, M.A., 2021. Frontiers and advances in green and sustainable inhibitors for corrosion applications: A critical review. *Journal of Molecular Liquids*, *321*, 114385.
37. Basik, M. and Mobin, M., 2022. Environmentally sustainable corrosion inhibitors in the oil and gas industry. In *Environmentally Sustainable Corrosion Inhibitors* (pp.405–421). Elsevier.
38. Kaur, J., Daksh, N. and Saxena, A., 2022. Corrosion inhibition applications of natural and eco-friendly corrosion inhibitors on steel in the acidic environment: an overview. *Arabian Journal for Science and Engineering*, *47*(1), pp.57–74.
39. Kobzar, Y.L. and Fatyeyeva, K., 2021. Ionic liquids as green and sustainable steel corrosion inhibitors: Recent developments. *Chemical Engineering Journal*, *425*, 131480.
40. Berdimurodov, E., Kholikov, A., Akbarov, K., Guo, L., Kaya, S., Katin, K.P., Verma, D.K., Rbaa, M., Dagdag, O. and Haldhar, R., 2022. Novel gossypol–indole modification as a green corrosion inhibitor for low-carbon steel in aggressive alkaline–saline solution. *Colloids and Surfaces A: Physicochemical and Engineering Aspects*, *637*, 128207.
41. Hrimla, M., Bahsis, L., Laamari, M.R., Julve, M. and Stiriba, S.E., 2022. An overview on the performance of 1, 2, 3-triazole derivatives as corrosion inhibitors for metal surfaces. *International Journal of Molecular Sciences*, *23*(1), p.16.
42. Rouifi, Z., Rbaa, M., Abousalem, A.S., Benhiba, F., Laabaissi, T., Oudda, H., Lakhrissi, B., Guenbour, A., Warad, I. and Zarrouk, A., 2020. Synthesis, characterization and corrosion inhibition potential of newly benzimidazole derivatives: Combining theoretical and experimental study. *Surfaces and Interfaces*, *18*, 100442.
43. Verma, C., Alrefaee, S.H., Rhee, K.Y., Quraishi, M.A. and Ebenso, E.E., 2021. Thiol (-SH) substituent as functional motif for effective corrosion protection: A review on current advancements and future directions. *Journal of Molecular Liquids*, *324*, 115111.
44. Ouakki, M., Galai, M. and Cherkaoui, M., 2022. Imidazole derivatives as efficient and potential class of corrosion inhibitors for metals and alloys in aqueous electrolytes: a review. *Journal of Molecular Liquids*, *345*, 117815.
45. Mehta, R.K., Gupta, S.K. and Yadav, M., 2022. Studies on pyrimidine derivative as green corrosion inhibitor in acidic environment: Electrochemical and computational approach. *Journal of Environmental Chemical Engineering*, *10*(5), 108499.
46. Yadav, P.K., Prakash, O., Ray, B. and Maiti, P., 2021. Functionalized polythiophene for corrosion inhibition and photovoltaic application. *Journal of Applied Polymer Science*, *138*(44), 51306.
47. Verma, C. and Quraishi, M.A., 2022. Green corrosion inhibitors derived through one-step multicomponent reactions: recent developments. *Environmentally Sustainable Corrosion Inhibitors*, pp.289–302.
48. Beniken, M., Salim, R., Ech–chihbi, E., Sfaira, M., Hammouti, B., Touhami, M.E., Mohsin, M.A. and Taleb, M., 2022. Adsorption behavior and corrosion inhibition mechanism of a polyacrylamide on C–steel in 0.5 M H_2SO_4: Electrochemical assessments and molecular dynamic simulation. *Journal of Molecular Liquids*, *348*, 118022.

49. Sanni, O., Iwarere, S.A. and Daramola, M.O., 2023. Investigation of eggshell agro-industrial waste as a potential corrosion inhibitor for mild steel in oil and gas industry. *Sustainability*, *15*(7), p.6155.
50. Benabdellah, M., Tounsi, A., Khaled, K.F. and Hammouti, B., 2011. Thermodynamic, chemical and electrochemical investigations of 2-mercapto benzimidazole as corrosion inhibitor for mild steel in hydrochloric acid solutions. *Arabian Journal of Chemistry*, *4*(1), pp.17–24.
51. El Ibrahimi, B., Jmiai, A., Bazzi, L. and El Issami, S., 2020. Amino acids and their derivatives as corrosion inhibitors for metals and alloys. *Arabian Journal of Chemistry*, *13*(1), pp.740–771.
52. Inzunza, R.G., Valdez, B. and Schorr, M., 2013. Corrosion inhibitor patents in industrial applications—a review. *Recent Patents on Corrosion Science*, *3*(2), pp.71–78.
53. Coelho, L.B., Lukaczynska-Anderson, M., Clerick, S., Buytaert, G., Lievens, S. and Terryn, H.A., 2022. Corrosion inhibition of AA6060 by silicate and phosphate in automotive organic additive technology coolants. *Corrosion Science*,*199*, 110188.
54. Lavoie, M., Baillie, J., Bourlon, E., O'Connell, E., MacKay, K., Boelens, I. and Risk, D., 2022. Sweet and sour: A quantitative analysis of methane emissions in contrasting Alberta, Canada, heavy oil developments. *Science of the Total Environment*, *807*, 150836.
55. Kuznetsov, Y.I., Andreev, N.N., Ibatullin, K.A. and Oleinik, S.V., 2002. Protection of low-carbon steel from carbon dioxide corrosion with volatile inhibitors. I. Liquid phase. *Protection of Metals*,*38*, pp.322–328.

4 A Journey from Traditional to Agricultural Waste as Corrosion Inhibitors

4.1 INTRODUCTION

Worldwide, there have been unprecedented challenges because of increasing pollution, which can primarily be attributed to fossil fuel consumption. This has not only caused global warming but has also been aggravated by the problem of the recent pandemic (COVID-19) [1–3]. These challenges have forced researchers to look for raw materials from benign sources in the form of renewable resources which are available in abundance. The processing of these renewable resources, specifically here we focus on agricultural wastes, which must be economically competitive compared to nonrenewable sources for their gainful utilization [4, 5]. These are economical renewable energy resources and are available in adequate quantity. Their accumulation in enormous amounts yearly causes environmental problems and, if not utilized, will lose substantially valuable resources. This has increased curiosity in conversion processes for utilizing agricultural waste for energy and power generation through direct and indirect conversion routes.

As the world is working toward Vision 2030, which aims to phase out toxic chemicals, corrosion scientists are intensifying research activities on green sources. Research activities have now been focused on natural inhibitors due to their availability, low cost, biodegradability, and eco-friendliness. Most of the work is devoted to studying food waste as a corrosion inhibitor; however, a mixture of waste formed by the mass market network is of particular interest. According to Food Waste Management Market: Global Industry Trends, Share, Size, Growth, Opportunity, and Forecast 2021–2026 [6–9], the mass market food wastebasket includes cereals, dairy products, fruits and vegetables, fish and seafood, processed food, and others. Most of the waste comprise vegetables and fruit, including the skins of a banana, tangerine/oranges, grapes, apples, carrots, avocados, tomatoes, skins of potatoes, carrots, and onions. In developing novel environmentally friendly corrosion inhibitors, in this chapter, we report the valorization of agricultural wastes as the source of "green" compounds. Valorization of agro-wastes can reduce their negative environmental impacts and has gained attention as a source of "green" compounds for developing sustainable chemistry that can minimize the impact profits on the environment and generate profits for the anti-corrosion protection area. World experience shows that currently there are several main areas of development of innovative technologies in

DOI: 10.1201/9781003441151-4

enhancing the corrosion resistance of metal surfaces – physicochemical modification of the surface by corrosion inhibitors based on "green" organic compounds [10–12]. Nevertheless, in both cases, there are several unresolved vital issues regarding forming functionalized protective films using "green" extract of raw materials. Thus, for all the undoubted advantages of using the flagships of "green chemistry"– agricultural waste extracts – their effectiveness is usually limited, and the film formation process is uncontrolled. This chapter will provide readers a clear understanding of the various vital aspects of agricultural waste technology and help them think and ponder the future research scope on agricultural residues to make this waste sustainable, efficient, and widely acceptable.

4.2 ECONOMIC AND INDUSTRIAL OPPORTUNITIES

Actives-controlled release of various formulations for agricultural purposes has various advantages such as economic, low dosages, minimal labor usage, safety, and low environmental impact. Agricultural wastes are exploited to develop agricultural actives, mostly as economically feasible products that may serve dual functions as fertilizer and carrier material.

4.3 AGRICULTURAL WASTES EXTRACT AS SUSTAINABLE CORROSION INHIBITORS

Worldwide, tonnes of agricultural waste are generated annually [13–16]. Out of this, a small amount is used, most of which is handled unscientifically and ends up in unmanaged waste sites, thus contributing to global warming potential. In some cases, agricultural residue is burned onsite, thereby causing great difficulties for the people not only in the surroundings but also in faraway places, giving rise to air pollution [17, 18]. Approximately one-third of all food produced today is estimated to go to landfill [19, 20]. It can be utilized as a corrosion inhibitor, which is biodegradable and eco-friendly.

4.3.1 Process of Extraction

One of the most extensive investigations in corrosion inhibition is the development of green corrosion inhibitors based on agricultural waste. Reports regarding using agricultural waste extracts in corrosion inhibition keep increasing yearly. The increasing number of publications proves this topic is essential in finding a better solution for corrosion problems by using agricultural waste extract-based corrosion inhibitors. The increasing number of investigations also indicates that green corrosion inhibitors from agricultural waste extract have great potential in corrosion prevention. The next section of this chapter focuses on agricultural waste extract-based corrosion inhibitors. Preparation, performance inhibition mechanism, and important aspects regarding the effect of agricultural waste in corrosion inhibition are revealed.

The preparation of agricultural waste extract usually starts with the drying process and is subsequently followed by grinding and sieving processes to transform it into powder form. Generally, the first drying process is compulsory for all parts except for juice extraction from the waste. The conventional drying process is typically performed at room temperature, which takes a long time. For instance, the drying process of the bark is approximately 20–30 days [21, 22], which the sun-drying process can do. After drying, various methods can isolate and extract the desired extract from the waste. Extraction methods include solvent extraction, distillation method, pressing, and sublimation according to the extraction principle [23–25]. In brief, extraction methods are based on heating, cooling, and separating the active compounds in the presence of the solvent [26]. Among these methods, solvent extraction is the most widely used method for waste extraction. Figure 4.1 presents the general solvent extraction procedures that can be used to extract agricultural waste. In the solvent extraction method, the solvent is required to diffuse into agricultural waste, solubilize, and finally extract the compounds [27, 28]. Figure 4.2 lists the examples of various extraction solvents and the resultant active extracts. Different solvents can be employed as an extraction medium for a different desired extract. Tan and Kassim [29] reported that ethanol extract showed exceptionally high tannin, total flavonoids, and total phenolics compared to acetone extract of *Rhizophora apiculata*. In the other study, 70% acetone was a more efficient solvent than either water or 80% methanol when each solvent was used to extract tannins from *Acacia nilotica* [30]. Most agricultural waste extracts are extracted by methanol, ethanol, or aqueous solvent as illustrated in Figure 4.2.

Factors such as solvent-to-solid ratio, type of solvents with varying polarities, extraction time, and temperature might affect the properties of extract via solvent extraction and the chemical composition and physical characteristics of the samples [31, 32]. For instance, increasing the solvent-to-solid ratio was found to work positively for enhancing phenol yields from black currant extraction [33–35]. However, the optimized value must be considered to balance the cost, solvent wastes, and avoidance of saturation effects. In addition, a smaller particle sample size can also increase the yield for extract. In addition, a combined system of aqueous and organic solvents can influence extract yield. For example, yield is significantly higher in an aqueous and organic solvent combination than using one system as the compound from agricultural waste extract is soluble in both types of systems. Mangrove tannins (*R. apiculata*) are more soluble in polar solvents than non-polar solvents. In a polar solvent, acetone can inhibit interactions between polyphenols and protein, and a combination of polar solvent solutions facilitates more polyphenolic compounds such as phenolic content, tannin, and flavonoid in the waste to be extracted. Microwave-assisted, enzyme-assisted, and ultrasound-assisted methods are other approaches used for agricultural waste extraction [36, 37]. Furthermore, methods based on compressed fluids as extracting agents like subcritical water, supercritical fluids, and pressurized fluids have also been used for extract preparation from agricultural waste. Supercritical fluids denote a new class of alternative solvents for preparing agricultural waste extracts, which allow selective separation of phytochemicals from the extracts at moderate temperatures and optimum processing time. Table 4.1 summarizes the benefits of these extraction methods. The other factor that needs to be considered for the agricultural waste extract preparation is the temperature during

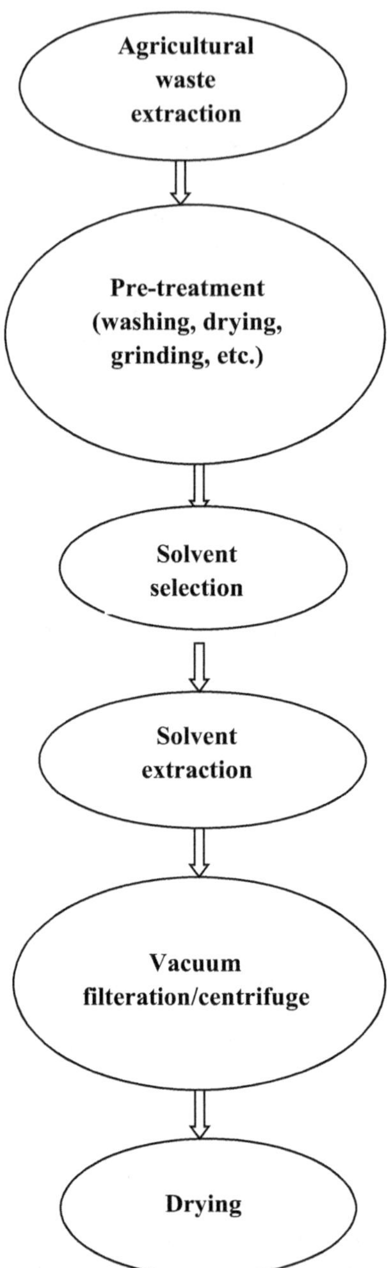

FIGURE 4.1 Flowchart of agricultural waste extraction process via solvent extraction.

the extraction process. A temperature range between 60°C and 80°C is preferable for the extraction process because the range can yield optimal extraction [38, 39]. The suitable temperature during extraction is chosen for effective solubility of the solvent extraction.

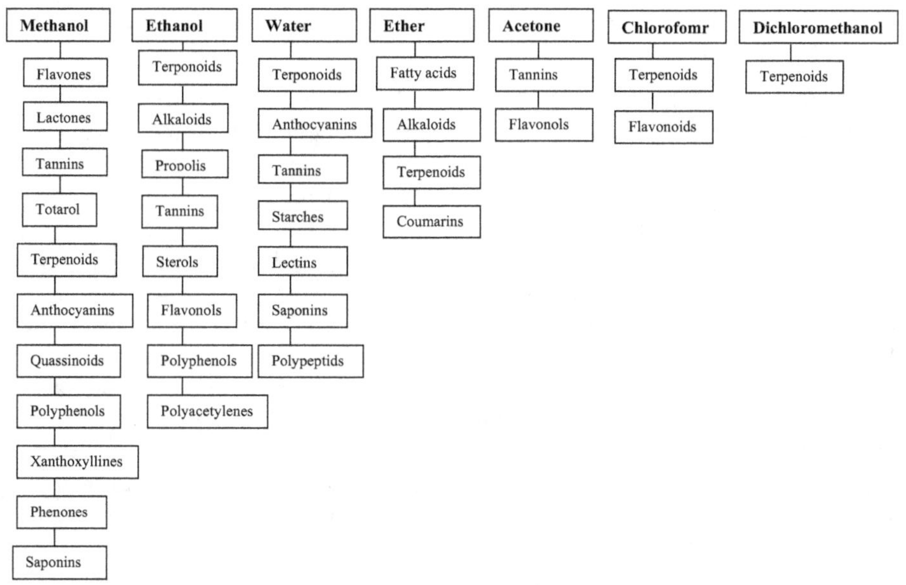

FIGURE 4.2 Various kinds of extracts from different solvents [41].

TABLE 4.1
Benefits of several extraction methods for agricultural waste extract

Solvent extraction	Low energy consumption, large production capacity, fast action, easy continuous operation, and ease of automation
Microwave irradiation	Microwave irradiation had a synergistic effect of breaking and heating which can increase reaction yields, reduce reaction times and avoid damage to active components that can occur during traditional, high-temperature heating methods
Enzyme-assisted extraction	A novel and effective way to release bounded compounds, allowing nearly whole waste matrix exploitation, and subsequently increase overall yield
Ultrasound-assisted extraction	Ultrasound energy facilitates organic and inorganic compounds leaching from waste matrix

Higher temperature causes higher solubility by increasing both solubility and mass transfer rates, decreasing the viscosity, and improving the solvent's movement in the matrices. In addition, the optimum temperature for the extraction process is compulsory without decomposing the active constituents (phytochemicals). High temperatures with a long time of extraction process can cause the oxidation of phenolic compounds, which results in a decrement in the extract.

4.3.2 Factors Influencing the Efficiency of Agricultural Wastes as Inhibitors

"Green chemistry" refers to an attempt to build a balanced approach to chemical risk management. This concept develops products that eliminate or reduce chemical pollution and waste management issues. Green chemistry is based on sustainability principles, reducing environmental consequences and preserving natural resources for future generations. Green corrosion inhibitors are also disposable and devoid of toxic metals and other potentially hazardous substances [41–43] and can prevent metal corrosion in different aggressive environments. They have many polar atoms and electron-rich linkages and are rich in organic molecules. Their molecules are adsorbed on the surface of metal or alloy during the inhibition phase and form a protective film. In the process, they may transfer an electron to the vacant d-orbitals of iron or metal atoms, resulting in coordinate bonds. Metal ions migrate into the electrolytic solution at the anode during corrosion and transport electrons from the metal to the cathode. Electron acceptors such as oxygen, oxidizing agents, or hydrogen ions are needed in the cathodic reaction. Eventually, they interact with either the reactions occurring at the anode, cathode, or both to reduce oxidation or reduction or both corrosive reactions [44, 45]. The factors influencing the efficiency of agricultural wastes as suitable inhibitors are listed below:

- Cost-effectiveness
- Availability
- Ability to oxidize the metal to form a protective layer on the metal/corrosive solution interface
- Presence of π-electrons and/or hetero atoms (S, P, N, O) with large structures that can donate lone pair electrons in the organic molecule
- Solubility
- Ability to adsorb on the metal surface

4.4 APPLICATIONS OF AGRICULTURAL WASTES AS ORGANIC GREEN CORROSION INHIBITORS

The corrosion process is a significant problem in industry, infrastructure, and technological advancement for sustainable development, resulting in premature failure of structural materials and increased maintenance costs. The inhibitive action of corrosion inhibitors occurs in two ways: (i) using environment modifiers to alter the corrosive media into a noncorrosive or less corrosive environment and (ii) by adsorbing green corrosion inhibitors onto the surface of metals to develop a protection layer of adsorbed inhibitors. A corrosion inhibitor is primarily concentrated on materials that are sustainable, readily available, harmless, low cost, environmentally friendly, and ecologically acceptable components. The quest for greener substitutes can be accomplished by modifying current products or exploring novel chemistry to reduce the toxicity and biodegradability of discharge materials.

Chemistry has raised much interest in the recent decade by customizing its chemicals and treatments to avoid pollutants and reduce waste. There was already a significant concern in the 1980s about waste generation and the use of hazardous and toxic chemicals. The Pollution and Prevention Act of the United States (1990) [46] provided the impetus for a watershed moment in environmental consciousness. Environmental challenges and their remedies have become a global priority. Several European and non-European countries, namely the UK, Italy, and Japan, have begun major green chemistry initiatives. Green chemistry has shown how basic scientific principles may protect public health and the environment cost-effectively through the years, covering areas such as polymeric materials, solvents, catalysis, technical modification, synthetic methodology advancement, and the prototype for environmentally friendly chemicals with vast wisdom of its impacts on the environment.

Agricultural waste materials are at the core of many industrial applications aligned with ideas of environment and sustainable chemistry. The demand for renewable resources has grown as nature supplies an immensely diverse spectrum of matrices with varying chemical compositions. As various types of agricultural waste are required for animal and human nutrition, their utilization for other objectives must be harmonized. Preferably, the agricultural waste that remains after the removal of the nutritional constituents, known as bio-waste, can be used as substrates for various green chemical purposes. Metal surface protection is one of the domains in which green chemistry is extensively employed in agricultural waste-derived products, delivering new ideas to reduce ecological consequences and wastes. Using agricultural waste as residues provides a cost-effective, environmentally friendly platform for manufacturing valuable chemicals as corrosion inhibitors.

4.4.1 Cooling Waters

To combat corrosion, a known method is the addition of synthetic corrosion and scale inhibitors in a recirculating cooling water system. In the work of Lohitkarn et al. [47] the authors utilized the extract from cassava leaf, agricultural waste in Thailand, as a green inhibitor. Corrosion inhibition efficiency using cassava leaf extract for mild carbon steel was investigated in a simulated cooling water solution under both stagnant and hydrodynamic conditions. In the case of static conditions, the inhibitor showed a corrosion inhibition efficiency of 97.4% with 800 ppm concentration at 35°C.

4.4.2 Concrete

Concrete is an artificial composite material with different components: water, fine and coarse aggregates (sand and gravel), and cement. It has gained popularity as a construction material around the globe, owing to its adaptability, economic issues, and durability. The primary motivations behind the creation of concrete structures are to reduce the emission of harmful gases such as carbon dioxide (CO_2) and other pollutants during the production of concrete; to make better use of waste products; to design and construct buildings that require less energy to operate over time; and to take advantage of concrete's thermal mass. The manufacturing process of cement leads to the release of a significant quantity of the greenhouse gas carbon dioxide. Finding alternatives to cement is becoming crucial. The ash that is produced when agricultural waste is burned as fuel presents a few problems, such as the contamination of the soil, but it also has the potential to be used as a cement substitute because of its pozzolanic qualities. Different agro-wastes have been utilized in the cement industry to improve concrete performance, including corn cob ash, rice husk ash, palm oil fuel ash, wood waste ash, bamboo leaf ash, coconut husk ash, groundnut shell ash, and sugarcane bagasse ash as substitutes for cement in concrete mixtures. The use of agricultural waste in cement and concrete could lead to a range of benefits which include reduced emissions of CO_2, and improved strength and durability properties of concrete resulting in lower concrete production costs.

The use of corrosion inhibitors can result in substantial cost savings on maintenance for reinforced concrete structures during their expected service life of 40–60 years. If the concrete is built with a low diffusion coefficient and adequate cover to minimize the rate at which chloride reaches the steel, the corrosion inhibitor may increase the period before corrosion occurs [48, 49]. A long-term improvement in the passivation state of steel reinforcement and other embedded steel in concrete structures is offered by corrosion-inhibiting admixtures, which are effective after the concrete has hardened. Common types of corrosion-inhibiting admixture include calcium nitrite, amino alcohols, and blending amino alcohols with organic inhibitors. Corrosion inhibitors derived from amino alcohols are added at a proportion of 3–4% by weight of the cement. The workability and air content of new concrete are marginally improved by a corrosion inhibitor based on calcium nitrite. The use of an amino alcohol-based inhibitor seems to decrease compressive strength while boosting workability. Corrosion inhibitors function in numerous ways, depending on their chemical composition. Corrosion inhibitors that contain calcium nitrite stabilize the passive layer on the steel surface and reduce its reactivity. The steel becomes inert once the layer is penetrated by chloride ions; therefore no reaction happens. Nitrites are anodic inhibitors because they hide the steel's anodic corrosion sites from chloride assault. Corrosion inhibitors based on amino alcohols form a monomolecular coating on the metal's surface, thus preventing chloride ions from reaching the encased steel. Moreover, they prevent the critical stage of corrosion where oxygen and water react at the cathodic sites on the steel.

Approximately400 million metric tons of agricultural waste are produced yearly in the world's developing nations. Waste products from many agricultural activities end up in landfills. Debris from agricultural production or human usage may be recycled into new products. Portland cement combinations have been examined to learn more about the features or attributes of agricultural waste which includes rice husk ash, sugarcane bagasse ash, sugarcane straw ash, palm oil fuel ash, maize cob, eucalyptus wood (bark), bamboo leaf ash, coconut husk ash, and cotton stalk. Pollution, landfilling, and expensive construction may be avoided if these wastes are recycled and used as a sustainable building material. To improve the durability and functionality of concrete, agro-waste seems to be a viable option.

4.4.3 Acid Pickling, Chemical Cleaning, and Boilers

Metals and alloys generally showed a high tendency toward corrosion because of acid. Metals corrode in acid solution as it can attack the metal surface to dissolve

TABLE 4.2
Summary of agricultural waste extract performance as corrosion inhibitor evaluated in different corrosive solution

Waste material	Extraction solvent	Substrate	Corrosive medium	Highest efficiency (%)	References
Maltaise orange peel	C_2H_5OH/H_2O	Carbon steel	HCl	95.00	[60]
Sugar cane waste	NaOH, HCl, H_2O	Iron plate	HCl	80.79	[61]
Pistachio nut hard shell	Water	Mild steel	HCl	92	[62]
Rollinia occidentalis seed	CH_3OH	Mild steel	HCl	79.7, 85.70, and 95.00	[63]
Lychee fruit waste (peel and seed)	C_2H_5OH	Mild steel	HCl	87.04, and 97.95	[64]
Green pea peel (*Pisum sativum*)	H_2O	Mild steel	HCl	84.00, 90.1, and 91.00	[65]
Elephant grass	Ethanol	Mild steel	HCl	95	[66]
Orange peel waste	H_2SO_4	Iron plate Mild steel	HCl	95.10, and 95.36	[67]
Solid tea waste	Deionized water	Boiler quality steel	HCl	84.53	[68]
Chinese gooseberry fruit shell	H_2O	Carbon steel	HCl	92.00, and 94.00	[69]
Cucumber (*Cucumis sativus*) peel	HCl	Carbon steel	HCl	80.34, and 82.70	[70]
Shrimp waste	Water	Carbon steel	HCl	95.1	[71]
Rice husk ash	NaOH	Carbon steel	NaCl	82	[72]
Hunteria umbellata seed husk extracts	H_2SO_4	Mild steel	H_2SO_4	92.00	[73]
Barley agriculture waste (maltbagasse)	H_2SO_4	Stainless steel	H_2SO_4	71.20, 92.80, and 97.10	[74]
Almond (*Prunus dulcis*) peel	H_2O	Mild steel	HCl	88.00	[75]
Human hair (HHR)	Methanol, ethanol, chloroform	Mild steel	HCl	85.09, 85.15, and 96.65	[76]

(*continued*)

TABLE 4.2 (Continued)
Summary of agricultural waste extract performance as corrosion inhibitor evaluated in different corrosive solution

Waste material	Extraction solvent	Substrate	Corrosive medium	Highest efficiency (%)	References
Longan seed and peel	H_2O	Mild steel	HCl	86.38, 92.35, and 92.93	[77]
Hen feather	H_2SO_4	Mild steel	H_2SO_4	95.5	[78]
Chlorella sorokiniana	BBM medium	Mild steel	HCl	94.6	[79]
Tomato peel waste	Ammonium oxalate and oxalic acid	Tin	NaCl and acetic acid	75.9	[80]
Chicken nail	Ethanol	Mild steel	H_2SO_4	74.04	[78]
Espresso spent ground coffee waste	Water	Carbon steel	HCl	94.83	[81]
Honeycomb waste	Ethanol	Stainless steel 304	H_2SO_4	97.29	[82]
Date palm waste	HCl	Carbon steel	HCl	89	[83]
Irvingia wombolu	Ethanol	Mild steel	HCl	97.87	[84]
Eggshell powder	Water	Stainless steel 316	NaCl	96.66	[85]
Waste *Allium capa* peel	Ethanol	X80 steel	HCl	95.8	[86]
Waste activated sludge	Water	Carbon steel	NaCl	80	[87]
Elaeis guineensis agricultural waste	Ethanol	Mild steel	HCl	93	[88]
Coconut husk fibers	Ethanol and acetone	Mild steel	H_2SO_4	79.9	[89]
Pomegranate peels	Ethanol	Mild steel	HCl	95	[90]
Vignaun guculata waste	Water	Aluminum	NaOH and H_2SO_4	79.63	[91]
Barley agro industrial waste	Water	Stainless steel 304	H_2SO_4	97	[74]
Banana peel	Water	Mild steel	H_2SO_4	87.44	[92]
Punica granatum	Methanol	Mild steel	HCl	80	[90]
Human hair extract	Water	Mild steel	HNO_3	98.58	[93]

it into its ions [50–52]; it occurs during several industrial processes to form stable corrosion products. For example, HCl and H_2SO_4 are mineral acids used in many industries for various metallurgical purposes like acid pickling, chemical cleaning, and descaling [53, 54], and mill scale removal from metal surfaces [55]. Meanwhile, in the oil and gas field, corrosion can be related to the nature of crude oil itself, which promotes corrosion due to its harmful impurities like naphthenic acid and sulfur [56, 57]. According to the literature, a lack of corrosion prevention caused a cost of corrosion-related issues such as maintenance, repair rehabilitation, and replacement of damaged structures. In South Africa, it is estimated that about ZAR166 billion is spent annually in South Africa in trying to combat or prevent corrosion problems. By applying NACE's 2016 figure of 3.4% of the global GDP as the cost of corrosion per annum, this means that, currently. In general, corrosion causes severe failure in some structures, such as rapid corrosion of metallic parts, which can be costly to repair the structure, leading to losing time during maintenance due to shutting down systems. Moreover, the corrosion repair process is hazardous and can be fatal and can cause injury to workers. These compel researchers to focus on corrosion as an essential issue that must be addressed aggressively worldwide.

The corrosion inhibition efficiency of many synthesized organic compounds in different acidic environments has been investigated from time to time. Still, the hazardous impacts of these inhibitors on marine and probably animal lives have yet not been fully understood. Therefore, alternative routes for corrosion inhibition, which could be biodegradable, non-toxic, environmentally acceptable, and economically feasible, have been searched for a few decades. Literature surveys on green corrosion inhibitors suggest that the essential requirements for selecting natural products include those containing phytochemicals or isolated organic compounds containing hetero atoms, and N, O, P, and S, having a π-electrons system [58, 59]. The inhibition efficiency of these compounds can be measured in terms of the number of mobile electrons present, the character of the orbital containing free electrons, and the density of electrons at hetero atoms.

4.5 CHALLENGE AND OUTLOOK

Several reports regarding agricultural waste as corrosion inhibitors have been published in the literature. However, there are also associated issues to consider. The drying process is the first step in preparing agricultural waste extract. Further advanced investigations into the drying process and other strategies, such as the optimum temperature and conditions, are required. This is a significant parameter in the extraction process. Due to the diverse, desirable agricultural waste qualities, multiple solvents were used throughout the solvent extraction process. A strong acidic or alkaline media is usually utilized, which harms health and the environment and raises the risk of hazardous waste. This will result in an issue regarding the safety regulations that must be followed and the expense of waste treatment, which is likely to rise. Consequently, green methods should be used to prepare agricultural waste-based corrosion inhibitors to avoid hazardous material discharges into the environment. A reproducibility test is required, and flaws detected throughout the analysis must be rectified. Finally, future studies should focus on the potential commercialization of agricultural waste extract as corrosion inhibitors.

4.6 SUMMARY

Although corrosion in metals and alloys is an ever-persistent and natural process, it poses several hazardous risks to human health and the environment. To suppress and mitigate the rate of the corrosion process, the most practical method is to employ a corrosion inhibitor as a preventative strategy. The use of agricultural waste extracts as a green corrosion inhibitor and as a replacement for conventional corrosion inhibitors has been well documented in the literature, and they have gained interest in recent years for use in various applications. Extracts can be made from several agricultural wastes, and the results revealed that these extracts have high efficiency in metal corrosion inhibition. The presence of heteroatoms (N, O, P, and S) and π-electrons in these extracts influences the efficiency of these corrosion inhibitors. Many experimental methodologies have confirmed agricultural waste extracts' superior efficiency as green corrosion inhibitors owing to their structural analysis and molecule surface interactions. Finally, the aggregation aspects that determine the characteristics and performance of agricultural waste-based corrosion inhibitors are intended to assist other researchers in better understanding the importance of evaluating agricultural waste-based green corrosion inhibitors in preventing metal corrosion.

REFERENCES

1. Mishra, J., Mishra, P. and Arora, N.K., 2021. Linkages between environmental issues and zoonotic diseases: with reference to COVID-19 pandemic. *Environmental Sustainability*, *4*(3), pp.455–467.
2. Yang, M., Chen, L., Msigwa, G., Tang, K.H.D. and Yap, P.S., 2022. Implications of COVID-19 on global environmental pollution and carbon emissions with strategies for sustainability in the COVID-19 era. *Science of the Total Environment*, *809*, 151657.
3. Oo, K.T. and Thin, M.M.Z., 2022. Climate change perspective: The advantage and disadvantage of COVID-19 pandemic. *Journal of Sustainability and Environmental Management*, *1*(2), pp.275–291.
4. Yadav, M., Goel, G., Hatton, F.L., Bhagat, M., Mehta, S.K., Mishra, R.K. and Bhojak, N., 2021. A review on biomass-derived materials and their applications as corrosion inhibitors, catalysts, food and drug delivery agents. *Current Research in Green and Sustainable Chemistry*, *4*, 100153.
5. Gautam, P., Upadhyay, S.N. and Dubey, S.K., 2020. Bio-methanol as a renewable fuel from waste biomass: current trends and future perspective. *Fuel*, *273*, 117783.
6. Boliko, M.C., 2019. FAO and the situation of food security and nutrition in the world. *Journal of Nutritional Science and Vitaminology*, *65*(Supplement), pp.S4–S8.
7. Papamichael, I., Chatziparaskeva, G., Pedreno, J.N., Voukkali, I., Candel, M.B.A. and Zorpas, A.A., 2022. Building a new mind set in tomorrow fashion development through circular strategy models in the framework of waste management. *Current Opinion in Green and Sustainable Chemistry*, *36*, 100638.
8. Al-Obadi, M., Ayad, H., Pokharel, S. and Ayari, M.A., 2022. Perspectives on food waste management: Prevention and social innovations. *Sustainable Production and Consumption*, *31*, pp.190–208.

9. Kharola, S., Ram, M., Mangla, S.K., Goyal, N., Nautiyal, O.P., Pant, D. and Kazancoglu, Y., 2022. Exploring the green waste management problem in food supply chains: A circular economy context. *Journal of Cleaner Production*, *351*, 131355.
10. Alnajjar, A.O., Abd El-Lateef, H.M., Khalaf, M.M. and Mohamed, I.M., 2022. Steel protection in acidified 3.5% NaCl by novel hybrid composite of CoCrO3/ polyaniline: Chemical fabrication, physicochemical properties, and corrosion inhibition performance. *Construction and Building Materials*, *317*, 125918.
11. Bello, M., Ochoa, N., Balsamo, V., López-Carrasquero, F., Coll, S., Monsalve, A. and González, G., 2010. Modified cassava starches as corrosion inhibitors of carbon steel: An electrochemical and morphological approach. *Carbohydrate Polymers*, *82*(3), pp.561–568.
12. Kuznetsov, Y.I., 2004. Physicochemical aspects of metal corrosion inhibition in aqueous solutions. *Russian Chemical Reviews*, *73*(1), p.75.
13. Alatzas, S., Moustakas, K., Malamis, D. and Vakalis, S., 2019. Biomass potential from agricultural waste for energetic utilization in Greece. *Energies*, *12*(6), p.1095.
14. Pawelczyk, A., 2005. EU Policy and Legislation on recycling of organic wastes to agriculture. *ISAH*, *1*, pp.64–71.
15. Pappu, A., Saxena, M. and Asolekar, S.R., 2007. Solid wastes generation in India and their recycling potential in building materials. *Building and Environment*, *42*(6), pp.2311–2320.
16. Maji, S., Dwivedi, D.H., Singh, N., Kishor, S. and Gond, M., 2020. Agricultural waste: Its impact on environment and management approaches. In: Bharagava, R. (ed.) *Emerging Eco-friendly Green Technologies for Wastewater Treatment* (vol. 18, pp.329–351). Springer.
17. Karić, N., Maia, A.S., Teodorović, A., Atanasova, N., Langergraber, G., Crini, G., Ribeiro, A.R. and Đolić, M., 2022. Bio-waste valorisation: Agricultural wastes as biosorbents for removal of (in) organic pollutants in wastewater treatment. *Chemical Engineering Journal Advances*, *9*, 100239.
18. Koul, B., Yakoob, M. and Shah, M.P., 2022. Agricultural waste management strategies for environmental sustainability. *Environmental Research*, *206*, 112285.
19. McLean, C., 2021. Zero waste: Less food to landfill for a more sustainable future. *Redress*, *30*(1), pp.22–25.
20. Kassem, N., Pecchi, M., Maag, A.R., Baratieri, M., Tester, J.W. and Goldfarb, J.L., 2022. Developing decision-making tools for food waste management via spatially explicit integration of experimental hydrothermal carbonization data and computational models using New York as a case study. *ACS Sustainable Chemistry & Engineering*, *10*(50), pp.16578–16587.
21. Salleh, S.Z., Yusoff, A.H., Zakaria, S.K., Taib, M.A.A., Seman, A.A., Masri, M.N., Mohamad, M., Mamat, S., Sobri, S.A., Ali, A. and Ter Teo, P., 2021. Plant extracts as green corrosion inhibitor for ferrous metal alloys: A review. *Journal of Cleaner Production*, *304*, 127030.
22. Muhammad, S.L. and Ibrahim, M.B., 2022. Adsorption, thermodynamic and kinetic studies of *Azadirachta indica* (AI) bark extract as a potential corrosion inhibitor for zinc in 0.25 M HNO_3 solution. *Applied Journal of Environmental Engineering Science*, *8*(4), pp.8–4.
23. Kumari, P., 2022. Plant extracts as corrosion inhibitors for aluminum alloy in NaCl environment –recent review. *Journal of the Chilean Chemical Society*, *67*(2), pp.5490–5495.
24. Thakur, A. and Kumar, A., 2021. Sustainable inhibitors for corrosion mitigation in aggressive corrosive media: a comprehensive study. *Journal of Bio-and Tribo-Corrosion*, *7*, pp.1–48.

25. Thakur, A., Assad, H., Kaya, S. and Kumar, A., 2022. Plant extracts as environmentally sustainable corrosion inhibitors II. In *Eco-Friendly Corrosion Inhibitors* (pp. 283–310). Elsevier.
26. Miralrio, A. and Espinoza Vázquez, A., 2020. Plant extracts as green corrosion inhibitors for different metal surfaces and corrosive media: A review. *Processes*, *8*(8), p.942.
27. Diaconeasa, Z., Iuhas, C.I., Ayvaz, H., Mortas, M., Farcaş, A., Mihai, M., Danciu, C. and Stanilă, A., 2022. Anthocyanins from agro-industrial food waste: Geographical approach and methods of recovery – A review. *Plants*, *12*(1), p.74.
28. Pagano, I., Campone, L., Celano, R., Piccinelli, A.L. and Rastrelli, L., 2021. Green non-conventional techniques for the extraction of polyphenols from agricultural food by-products: A review. *Journal of Chromatography A*, *1651*, 462295.
29. Tan, K.W. and Kassim, M.J., 2011. A correlation study on the phenolic profiles and corrosion inhibition properties of mangrove tannins (*Rhizophora apiculata*) as affected by extraction solvents. *Corrosion Science*, *53*(2), pp.569–574.
30. Ishak, C.Y. and Elgailani, I.E.H., 2016. Methods for extraction and characterization of tannins from some *Acacia* species of Sudan. *Pakistan Journal of Analytical & Environmental Chemistry*, *17*(1), p.7.
31. da Silva, R.F., Carneiro, C.N., de Sousa, C.B.D.C., Gomez, F.J., Espino, M., Boiteux, J., Fernández, M.D.L.Á., Silva, M.F. and Dias, F.D.S., 2022. Sustainable extraction bioactive compounds procedures in medicinal plants based on the principles of green analytical chemistry: A review. *Microchemical Journal*, *175*, 107184.
32. Jha, A.K. and Sit, N., 2022. Extraction of bioactive compounds from plant materials using combination of various novel methods: A review. *Trends in Food Science & Technology*, *119*, pp.579–591.
33. Pap, N., Beszédes, S., Pongrácz, E., Myllykoski, L., Gábor, M., Gyimes, E., Hodúr, C. and Keiski, R.L., 2013. Microwave-assisted extraction of anthocyanins from black currant marc. *Food and Bioprocess Technology*, *6*, pp.2666–2674.
34. Pap, N., Beszédes, S., Pongrácz, E., Myllykoski, L., Gábor, M., Gyimes, E., Hodúr, C. and Keiski, R.L., 2013. Microwave-assisted extraction of anthocyanins from black currant marc. *Food and Bioprocess Technology*, *6*, pp.2666–2674.
35. Pinelo, M., Rubilar, M., Jerez, M., Sineiro, J. and Núñez, M.J., 2005. Effect of solvent, temperature, and solvent-to-solid ratio on the total phenolic content and antiradical activity of extracts from different components of grape pomace. *Journal of Agricultural and Food Chemistry*, *53*(6), pp.2111–2117.
36. Marić, M., Grassino, A.N., Zhu, Z., Barba, F.J., Brnčić, M. and Brnčić, S.R., 2018. An overview of the traditional and innovative approaches for pectin extraction from plant food wastes and by-products: Ultrasound-, microwaves-, and enzyme-assisted extraction. *Trends in Food Science & Technology*, *76*, pp.28–37.
37. Panja, P., 2018. Green extraction methods of food polyphenols from vegetable materials. *Current Opinion in Food Science*, *23*, pp.173–182.
38. Oh, S.H., Ahn, J., Kang, D.H. and Lee, H.Y., 2011. The effect of ultrasonificated extracts of Spirulina maxima on the anticancer activity. *Marine Biotechnology*, *13*, pp.205–214.
39. Hifney, A.F., Fawzy, M.A., Abdel-Gawad, K.M. and Gomaa, M., 2016. Industrial optimization of fucoidan extraction from *Sargassum* sp. and its potential antioxidant and emulsifying activities. *Food Hydrocolloids*, *54*, pp.77–88.
40. Cowan, M.M., 1999. Plant products as antimicrobial agents. *Clinical Microbiology Reviews*, *12*(4), pp.564–582.

41. Li, E., Liu, S., Luo, F. and Yao, P., 2023. Amino acid imidazole ionic liquids as green corrosion inhibitors for mild steel in neutral media: Synthesis, electrochemistry, surface analysis and theoretical calculations. *Journal of Electroanalytical Chemistry, 944*, 117650.
42. Benzidia, B., Barbouchi, M., Hsissou, R., Zouarhi, M., Erramli, H. and Hajjaji, N., 2022. A combined experimental and theoretical study of green corrosion inhibition of bronze B66 in 3% NaCl solution by *Aloe saponaria* (syn. *Aloe maculata*) tannin extract. *Current Research in Green and Sustainable Chemistry, 5*, 100299.
43. Shahryari, Z. and Gheisari, K., 2023. Corrosion inhibition properties of *Ziziphus spina-christi* leaves extract and zinc cations on mild steel in sodium chloride solution. *Materials Chemistry and Physics, 299*, 127474.
44. Ismail, N., Mujad, S.M., Zulkifli, M.F.R., Izionworu, V.O., Ghazali, M.J. and Nik, W.M.N.W., 2022. A review on application of marine algae as green corrosion inhibitors in acid medium. *Vietnam Journal of Chemistry, 60*(4), pp.409–416.
45. Barman, B. and Banjare, M.K., 2023. Deep understanding of corrosion inhibition mechanism based on first-principle calculations. In *Computational Modelling and Simulations for Designing of Corrosion Inhibitors* (pp.55–78). Elsevier.
46. Johnson, S.M., 1992. From reaction to proaction: the 1990 pollution prevention act. *Columbia Journal of Environmental Law, 17*, p.153.
47. Lohitkarn, J., Hemwech, P., Chantiwas, R. and Jariyaboon, M., 2021. The role of cassava leaf extract as green inhibitor for controlling corrosion and scale problems in cooling water systems. *Journal of Failure Analysis and Prevention, 21*, pp.847–860.
48. Tiwari, A.K., Goyal, S. and Luxami, V., 2023, February. Influence of corrosion inhibitors on two different concrete systems under combined chloride and carbonated environment. *Structures, 48*, pp.717–735.
49. Naderi, R., Bautista, A., Shagñay, S. and Velasco, F., 2023. Licorice (*Glycyrrhiza glabra*) as corrosion inhibitor of carbon steel reinforcing bars in mortar and its synergic effect with nitrite. *Journal of Industrial and Engineering Chemistry, 129*, pp.620–633.
50. Kahkesh, H. and Zargar, B., 2023. Estimating the anti-corrosive potency of 3-nitrophthalic acid as a novel and natural organic inhibitor on corrosion monitoring of mild steel in 1M HCl solution. *Inorganic Chemistry Communications, 158*, 111533.
51. Abbas, A., Adesina, A.Y. and Suleiman, R.K., 2023. Influence of organic acids and related organic compounds on corrosion behavior of stainless steel – A critical review. *Metals, 13*(8), p.1479.
52. Toghan, A., Fawzy, A., Alakhras, A.I., Sanad, M.M., Khairy, M. and Farag, A.A., 2023. Correlating experimental with theoretical studies for a new ionic liquid for inhibiting corrosion of carbon steel during oil well acidification. *Metals, 13*(5), p.862.
53. Agarwal, C. and Pandey, A.K., 2023. Remediation and recycling of inorganic acids and their green alternatives for the sustainable industrial chemical processes. *Environmental Science: Advances, 2*, pp.1306–1339.
54. Muhsan, M.A. and Zahoor, A.F., 2023. Reclamation of mineral acids from various waste streams using solvent extraction technique: A review. *Geosystem Engineering, 26*(5), pp.218–238
55. Enekwe, C.B., Igwilo, C.N., Ude, C. and Elendu, C.C., 2023. Optimization and effect of inhibitive action of *Vernonia amygdalina* and *Azadirachta indica* leave extracts on corrosion of mild steel in acidic medium: An innovative approach. *Ebonyi State College of Education, Ikwo Journal of Educational Research, 9*(1), pp.1–44.
56. Sotoudeh, Y., Niksokhan, M.H., Karbassi, A. and Sarafrazi, M.R., 2023. Review on naphthenic acids: An important environmental pollutants caused by oil extraction and industries. *Pollution, 9*(1), pp.254–270.

57. Yuan, Y., Gao, X., Zhou, J., Liu, G., Kuang, X., Yang, L. and Liao, R., 2022. A review: Research on corrosive sulphur in electrical power equipment. *High Voltage*, *7*(2), pp.209–221.
58. Donkor, S., Song, Z., Jiang, L. and Chu, H., 2022. An overview of computational and theoretical studies on analyzing adsorption performance of phytochemicals as metal corrosion inhibitors. *Journal of Molecular Liquids*, *359*, 119260.
59. Wei, H., Heidarshenas, B., Zhou, L., Hussain, G., Li, Q. and Ostrikov, K.K., 2020. Green inhibitors for steel corrosion in acidic environment: State of art. *Materials Today Sustainability*, *10*, 100044.
60. M'hiri, N., Veys-Renaux, D., Rocca, E., Ioannou, I., Boudhrioua, N.M. and Ghoul, M., 2016. Corrosion inhibition of carbon steel in acidic medium by orange peel extract and its main antioxidant compounds. *Corrosion Science*, *102*, pp.55–62.
61. Rahayu, P.P., Sundari, C.D.D. and Farida, I., 2018, November. Corrosion inhibition using lignin of sugarcane bagasse. IOP Conference Series: Materials Science and Engineering, *434*(1), 012087.
62. Shahmoradi, A.R., Talebibahmanbigloo, N., Javidparvar, A.A., Bahlakeh, G. and Ramezanzadeh, B., 2020. Studying the adsorption/inhibition impact of the cellulose and lignin compounds extracted from agricultural waste on the mild steel corrosion in HCl solution. *Journal of Molecular Liquids*, *304*, 112751.
63. Alvarez, P.E., Fiori-Bimbi, M.V., Neske, A., Brandán, S.A. and Gervasi, C.A., 2018. *Rollinia occidentalis* extract as green corrosion inhibitor for carbon steel in HCl solution. *Journal of Industrial and Engineering Chemistry*, *58*, pp.92–99.
64. Liao, L.L., Mo, S., Luo, H.Q. and Li, N.B., 2018. Corrosion protection for mild steel by extract from the waste of lychee fruit in HCl solution: Experimental and theoretical studies. *Journal of Colloid and Interface Science*, *520*, pp.41–49.
65. Srivastava, M., Tiwari, P., Srivastava, S.K., Kumar, A., Ji, G. and Prakash, R., 2018. Low cost aqueous extract of *Pisum sativum* peels for inhibition of mild steel corrosion. *Journal of Molecular Liquids*, *254*, pp.357–368.
66. Alaneme, K.K., Olusegun, S.J. and Alo, A.W., 2016. Corrosion inhibitory properties of elephant grass (*Pennisetum purpureum*) extract: Effect on mild steel corrosion in 1 M HCl solution. *Alexandria Engineering Journal*, *55*(2), pp.1069–1076.
67. Xavier Stango, S.A. and Vijayalakshmi, U., 2018. Studies on corrosion inhibitory effect and adsorption behavior of waste materials on mild steel in acidic medium. *Journal of Asian Ceramic Societies*, *6*(1), pp.20–29.
68. Pal, A. and Das, C., 2020. A novel use of solid waste extract from tea factory as corrosion inhibitor in acidic media on boiler quality steel. *Industrial Crops and Products*, *151*, 112468.
69. Dehghani, A., Bahlakeh, G. and Ramezanzadeh, B., 2019. A detailed electrochemical/theoretical exploration of the aqueous Chinese gooseberry fruit shell extract as a green and cheap corrosion inhibitor for mild steel in acidic solution. *Journal of Molecular Liquids*, *282*, pp.366–384.
70. Al-Senani, G.M., 2016. Corrosion inhibition of carbon steel in acidic chloride medium by *Cucumis sativus* (cucumber) peel extract. *International Journal of Electrochemical Science*, *11*(1), pp.291–302.
71. Farag, A.A., Ismail, A.S. and Migahed, M.A., 2018. Environmental-friendly shrimp waste protein corrosion inhibitor for carbon steel in 1 M HCl solution. *Egyptian Journal of Petroleum*, *27*(4), pp.1187–1194.
72. Fakir, S.H., Kumar, K.C., Ali, H., Rawat, J. and Farooqi, I.H., 2023. Synthesis and evaluation of green corrosion inhibitor from rice husk to mitigate corrosion of carbon steel in NaCl environment. *Materials Today: Proceedings*, *82*, pp.29–37.

73. Alaneme, K.K., Olusegun, S.J. and Adelowo, O.T., 2016. Corrosion inhibition and adsorption mechanism studies of *Hunteria umbellata* seed husk extracts on mild steel immersed in acidic solutions. *Alexandria Engineering Journal*, *55*(1), pp.673–681.
74. Matos, L.A.C., Taborda, M.C., Alves, G.J.T., da Cunha, M.T., do Prado Banczek, E., de Fátima Oliveira, M., D'Elia, E. and Rodrigues, P.R.P., 2018. Application of an acid extract of barley agro-industrial waste as a corrosion inhibitor for stainless steel AISI 304 in H_2SO_4. *International Journal of Electrochemical Science*, *13*(2), pp.1577–1593.
75. Pal, S., Lgaz, H., Tiwari, P., Chung, I.M., Ji, G. and Prakash, R., 2019. Experimental and theoretical investigation of aqueous and methanolic extracts of *Prunus dulcis* peels as green corrosion inhibitors of mild steel in aggressive chloride media. *Journal of Molecular Liquids*, *276*, pp.347–361.
76. Hemapriya, V., Prabakaran, M., Chitra, S., Swathika, M., Kim, S.H. and Chung, I.M., 2020. Utilization of biowaste as an eco-friendly biodegradable corrosion inhibitor for mild steel in 1 mol/L HCl solution. *Arabian Journal of Chemistry*, *13*(12), pp.8684–8696.
77. Liao, L.L., Mo, S., Luo, H.Q. and Li, N.B., 2017. Longan seed and peel as environmentally friendly corrosion inhibitor for mild steel in acid solution: Experimental and theoretical studies. *Journal of Colloid and Interface Science*, *499*, pp.110–119.
78. Olawale, O., Bello, J.O., Ogunsemi, B.T., Uchella, U.C., Oluyori, A.P. and Oladejo, N.K., 2019. Optimization of chicken nail extracts as corrosion inhibitor on mild steel in 2M H_2SO_4. *Heliyon*, *5*(11).
79. de Oliveira, G.A., Teixeira, V.M., da Cunha, J.N., dos Santos, M.R., Paiva, V.M., Rezende, M.J.C., do Valle, A.F. and D'Elia, E., 2021. Biomass of microalgae *Chlorella sorokiniana* as green corrosion inhibitor for mild steel in HCl solution. *International Journal of Electrochemical Science*, *16*(2), 210249.
80. Halambek, J., Cindrić, I. and Grassino, A.N., 2020. Evaluation of pectin isolated from tomato peel waste as natural tin corrosion inhibitor in sodium chloride/acetic acid solution. *Carbohydrate Polymers*, *234*, 115940.
81. da Costa, M.A., Gois, J.S.D., Toaldo, I.M., Bauerfeldt, A.C.F., Batista, D.B., Bordignon-Luiz, M.T., do Lago, D.C., Luna, A.S. and Senna, L.F.D., 2020. Optimization of espresso spent ground coffee waste extract preparation and the influence of its chemical composition as an eco-friendly corrosion inhibitor for carbon steel in acid medium. *Materials Research*, *23*(5).
82. Gapsari, F., Madurani, K.A., Simanjuntak, F.M., Andoko, A., Wijaya, H. and Kurniawan, F., 2019. Corrosion inhibition of honeycomb waste extracts for 304 stainless steel in sulfuric acid solution. *Materials*, *12*(13), p.2120.
83. Al-Senani, G.M. and Alshabanat, M., 2018. Study the corrosion inhibition of carbon steel in 1 M HCl using extracts of date palm waste. *International Journal of Electrochemical Science*, *13*, pp.3777–3788.
84. Mbah, C.N., Onah, C.C. and Nnakwo, K.C., 2020. Effectiveness of *Irvingia wombolu* extract on corrosion inhibition of mild steel in hydrochloric acid solution. *Engineering Research Express*, *2*(1), 015039.
85. Sanni, O. and Popoola, A.P.I., 2019. Assessment of concentration, temperature and exposure time effect on waste product as a sustainable inhibitor for stainless steel corrosion: optimization using response surface method. *Journal of Bio-and Tribo-Corrosion*, *5*, pp.1–14.
86. Ituen, E., Yuanhua, L., Singh, A. and Li, R., 2021. Chemical modification of waste *Allium cepa* peels to Cu-complex composite and application as eco environmental oilfield anticorrosion additive. *Journal of King Saud University – Engineering Sciences*, *33*(6), pp.375–385.

87. Go, L.C., Holmes, W., Depan, D. and Hernandez, R., 2019. Evaluation of extracellular polymeric substances extracted from waste activated sludge as a renewable corrosion inhibitor. *PeerJ*, *7*, e7193.
88. Hussin, M.H., Rahim, A.A., Ibrahim, M.N.M. and Brosse, N., 2015. Improved corrosion inhibition of mild steel by chemically modified lignin polymers from *Elaeis guineensis* agricultural waste. *Materials Chemistry and Physics*, *163*, pp.201–212.
89. Umoren, S.A., Solomon, M.M., Eduok, U.M., Obot, I.B. and Israel, A.U., 2014. Inhibition of mild steel corrosion in H_2SO_4 solution by coconut coir dust extract obtained from different solvent systems and synergistic effect of iodide ions: Ethanol and acetone extracts. *Journal of Environmental Chemical Engineering*, *2*(2), pp.1048–1060.
90. Ashassi-Sorkhabi, H., Mirzaee, S., Rostamikia, T. and Bagheri, R., 2015. Pomegranate (*Punica granatum*) peel extract as a green corrosion inhibitor for mild steel in hydrochloric acid solution. *International Journal of Corrosion*, 2015(1), 197587.
91. Umoren, S.A., Obot, I.B., Akpabio, L.E. and Etuk, S.E., 2008. Adsorption and corrosive inhibitive properties of *Vignaun guiculata* in alkaline and acidic media. *Pigment & Resin Technology*, *37*(2), pp.98–105.
92. Manikandan, C.B., Balamurugan, S., Balamurugan, P. and Beneston, S.L., 2019. Corrosion inhibition of mild steel by using banana peel extract. *International Journal of Innovative Technology and Exploring Engineering*, *8*, pp.1372–1375.
93. Sathiyapriya, T. and Rathika, G., 2015. Corrosion inhibition efficiency of bio waste on mild steel in acid media. *Oriental Journal of Chemistry*, *31*, pp.1703–1710.

5 Modern Testing and Analyzing Techniques in Corrosion Inhibition

5.1 INTRODUCTION

According to the International Union of Pure and Applied Chemistry (IUPAC), corrosion is "an irreversible interfacial reaction of a material (metal, ceramic, or polymer) with its environment that results in consumption of the material or the dissolution into the material of a component of the environment". Therefore, engineers are primarily concerned with metal corrosion. Corrosion inhibition is a billion-dollar industry, as corrosion affects almost all industries. Several methods are employed in testing corrosion inhibition. The methods for testing corrosion inhibitors include gravimetric or weight loss methods, volumetric methods, such as ultrasonic inspection and electric resistance (ER) probe monitoring, and standard electrochemical techniques, such as linear polarization resistance and potentiodynamic measurements. Other methods include surface profilometry (laser and optical microscopy) for localized corrosion assessment, bubble test, wheel test, and low pressure and high pressure. The detection and analysis of corrosion products is an essential aspect of a failure analysis to determine the underlying cause and the corrective action needed to avert future occurrences. However, the testing of any sample is only as good as the sample being examined, because many oxide and sulfide corrosion products degrade rapidly when exposed to air.

Corrosion monitoring involves gathering corrosion rates in a real-world industrial scenario, which can subsequently be turned into valuable data for a corrosion management strategy. The intricacy of corrosion monitoring becomes completely notable when one realizes that there are numerous types of corrosion (uniform or localized), corrosion rates vary over distances, and no solitary corrosion monitoring method is designed to measure corrosion rate under diverse circumstances [1, 2]. Continuous corrosion monitoring can aid in detecting corrosion impairment over time and the provision to take immediate measures before the destruction becomes uncontrollable. Some industries are stunning in neglecting corrosion regulators until a corrosion-related event happens. Corrosion scrutiny is used in several industries for routine inspections, repairs, and maintenance [3–5]. Several methods to investigate the corrosion inhibition action include gravimetric, electrochemical, and surface analysis. A conventional method, such as weight loss measurement (WLM), is the most

DOI: 10.1201/9781003441151-5

common method used to determine corrosion inhibitor performance from agricultural waste extracts, and an inhibition mechanism can be proposed.

Meanwhile, modern measurements of electrochemical processes such as potentiodynamic polarization (PDP) study, electrochemical impedance spectroscopy (EIS) [6–8], or electrochemical noise measurement are introduced. These time-saving measurements have produced comparative results to the conventional gravimetric method [9, 10]. The kinetics information of the corrosion in the presence of an inhibitor from the electrochemical process can eloquent the inhibition mechanism. Although these methods differ theoretically, the trend of the results is in magnitude that indicates results are acceptable and can be used to support the corrosion inhibition efficiency.

5.2 ANALYSIS IN CORROSION

Testing and assessing materials for corrosion resistance in various settings is known as corrosion analysis. In industrial and infrastructural applications, corrosion analysis can aid in corrosion problem resolution, prevention, or mitigation [11, 12]. The elements and varieties of corrosion products can be determined through corrosion analysis using various analytical methods, such as scanning electron microscopy (SEM) or X-ray diffraction (XRD). SEM is frequently used to identify the elements present in the corrosion products [13]. However, numerous additional experimental techniques, including XRD, can be used to offer critical information. The most popular approach for obtaining chemical identification of corrosion products is energy-dispersive X-ray spectroscopy (EDX) in conjunction with an SEM. EDX has a detectability limit of approximately 0.1 weight percent for elements C through U [14, 15]. The scanned electron beam reacts with the material's surface, producing secondary and backscattered electrons and X-rays. These X-rays are unique to the element that makes them and can be used to determine the elemental makeup of corrosion products. Unless standards are used, the analysis generates semiquantitative results depending on the program and presumed instrumental parameters. The corrosion product on a metal or fracture surface is frequently the initial EDX examination. This can reveal which metals (C, O, F, S, or Cl) are present, but the specific composition of the corrosion product cannot be determined. While the studied X-rays originate on the surface, they are released from an excluded volume that can be several microns deep. This excluded volume can contain the base metal or numerous corrosion scale layers.

Scale removal from the base metal or fracture surface can enable a more accurate analysis of the corrosion product chemistry. Using acetone-softened cellulose acetate mimicking tape to eliminate scale from the base metal is a good approach. When using the tape process, an unused piece should be assessed to detect probable contaminating elements from the tape that are not in the scale. Furthermore, if a C or Au coating is employed to give electrical conductivity, the low-energy X-rays are inhibited, and any data reduction is skewed. Another challenge with EDX analysis is a high enough accelerating voltage to excite the necessary X-rays. Furthermore, peak overlap can make understanding data challenging and necessitate dependence on software for discriminating; for example, the Mo-La and S-Ka peaks directly overlap, making determining the existence of one or both difficult. To confirm the presence of Mo and/or S, increase the accelerating voltage to include the Mo-Ka or utilize a wavelength dispersive X-ray spectrometer.

The optimum method for determining element position is a metallographic cross-sectional study or a focused ion beam-produced section with subsequent EDX analysis. Gravimetric evaluation/weight loss method, electrochemical analysis, and surface analysis are some of the approaches used to monitor and analyze corrosion inhibition activity.

5.2.1 Weight Loss Measurement

The most popular way to determine the efficacy of corrosion inhibitors based on agricultural waste extracts is to utilize a traditional method recognized as WLM.

In the meantime, advanced electrochemical measurement techniques like EIS, PDP, and harmonic distortion analysis have been introduced. Such time-saving measures have yielded specific results comparable to those obtained using the traditional gravimetric method [16–18]. The open-circuit potential is used initially to acquire a consistent current measurement on the metal surface. Although the methodologies differ theoretically, the results are similar in magnitude, demonstrating that the results are satisfactory and might be utilized to enhance corrosion inhibition efficiency. The inhibitor's effectiveness from WLM is calculated based on substrate weight loss [19, 20]. The substrate weights are recorded before and after the corrosive medium immersion. Then, the corrosion resistance of substrates is evaluated based on the measurements of corrosion rate and inhibition efficiency as shown in the following equations:

$$W = W_{initial} - W_{final} \tag{5.1}$$

where W_{initial} indicates the weight before the immersion test and W_{final} denotes weight following immersion. While equation (5.2) was used to obtain the corrosion rate in the absence and presence of inhibitors

$$Corrosionrate = 87.6W/DAT \tag{5.2}$$

where W denotes loss in weight (mg) after exposure time T (hours), A is the surface area of the specimen (cm^2) and D is the density (g/cm^3) of the substrate.

By conducting a weight loss experiment, Oliveira et al. [21] investigated the efficiency of microalgae *Chlorella sorokiniana* as a sustainable corrosion inhibitor for mild steel in 1 mol L^{-1} HCl solution. Utilizing 100 mg L^{-1} of microalgae *C. sorokiniana* and a 24-hour immersion time, the inhibition efficiency of 94.6% was achieved. The corrosion rate was obtained using the equation mentioned above. As the inhibitor concentrations increase, the corrosion rate reduces. The adsorption of the inhibitor followed the Langmuir isotherm, which admits the formation of a monolayer of the inhibitor molecules on the metal surface. The inhibitor molecules explain this discovery of increased coverage of the metal surface. This relationship suggests that the essential components of the microalgae *C. sorokiniana* are stable on the base metal, implying that these compounds revealed a significant interaction with the metallic surface. Sanni et al. [22] examined the corrosion inhibitive performance of palm kernel shell extract as on stainless steel in a seawater environment. The authors reported 90% efficiency of this inhibitor in the seawater solution, which is efficient for practical application. Further, the authors confirmed that the palm kernel shell extract was attached firmly to the metal surface by forming a protective film over the stainless steel surface.

5.2.2 Electrochemical Testing in Corrosion

The voltage and current of oxidation/reduction reactions are measured and/or controlled during electrochemical corrosion testing. Manipulation and measurement of these two variables allow for a variety of experiments. Most experiments include applying a voltage to the working electrode and measuring the current that results. American Society for Testing and Materials/ National Association of Corrosion Engineers (ASTM/NACE) methods, cyclic polarization, impedance spectroscopy, linear polarization tests, electrochemical testing for hydrogen permeation, real-time corrosion forecasts for materials, and others are among the electrochemical techniques.

5.2.2.1 Potentiodynamic Polarization

Potentiodynamic polarization (PDP) measurement with various temperatures in the absence and presence of different inhibitor concentrations during certain immersion times is frequently used. This measurement uses a three-electrode cell with working, counter, and reference electrodes [23–25]. Different types of electrode materials can be used for various functions. For instance, a platinum electrode was used as a counter electrode, a saturated calomel electrode as the reference electrode, and the cylindrical rod of C-steel alloy was used as a working electrode [26]. Based on the current response as a function of applied potential, several changes, such as the shape of polarization curves and the position of anodic or cathodic current densities compared to the curve without the presence of a corrosion inhibitor, can be used to monitor the efficiency of corrosion inhibitors in corrosion inhibition. Different effects of corrosion inhibitors on corrosion inhibition efficiency can be investigated by varying the concentration of corrosion inhibitors, tested temperatures, and different agricultural waste. Sahin et al. [27] compared the polarization curve of mild steel electrodes in the presence and absence of *Phoenix dactylifera* (date palm) seed extract after 6 hours of immersion in 1 M HCl solution. The authors found that the shape of the curve with and without inhibitor is unchanged, suggesting the corrosion mechanism that occurs on the surface of mild steel is similar. However, the anodic or cathodic current densities are lowered compared to the current density in the absence of an inhibitor on the mild steel surface. These findings have proven that the inhibitor inhibits corrosion through the rise of inhibition protection layer formation assembled on the metal surface.

In the other works, an apparent effect of different concentrations of *Heterophragma adenophyllum* extract on the corrosion behavior of Fe–C steel in 0.5 M HCl as reported by Pahuja et al. [28]. They have also interpreted that the shape of the polarization curve indicates *H. adenophyllum* extract corrosion inhibition efficiency. *H. adenophyllum* can function as an efficient corrosion inhibitor whereby the anodic–cathodic current densities are shifted to lower values at a higher *H. adenophyllum* extract concentration. This finding indicated that the *H. adenophyllum* extract could resist anodic and cathodic branch reactions. In addition, the corrosion inhibition efficiency of *H. adenophyllum* extract is also affected by temperature. The authors reported that a higher temperature would decrease the inhibition efficiency of *H. adenophyllum* extract due to desorption of the protective layer's constituent molecules from the metal surface. In addition, the PDP method is also used to calculate corrosion potential, corrosion current density, corrosion rate, and inhibition efficiency by exploiting the polarization curves. For example, corrosion potential and corrosion current density were calculated from the intersection of anodic and cathodic Tafel slopes of the polarization curves using CorrWare software or NOVA software [28–30]. Meanwhile, corrosion rate and inhibition efficiency from PDP measurement can be calculated as shown in equations (5.3) and (5.4) [31], respectively:

$$Corrosion\ rate\left(V_{COR}, g / m^2\right) = \frac{i_{COR} xtxM}{F} \tag{5.3}$$

$$Inhibition\ efficiency(\%) = \frac{\left(i_{CORR}\right)_a - \left(i_{CORR}\right)_b}{\left(i_{CORR}\right)_a} x100 \tag{5.4}$$

where i_{COR} is current density (Am^{-2}), M denotes the molar weight of the substrate ($gmol^{-1}$), F represents Faraday constant ($Cmol^{-1}$), t is the exposure time (s), $(i_{COR})_a$ signifies corrosion current density in the absence of the inhibitor (Am^{-2}), and $(i_{COR})_b$ denotes corrosion current density in the presence of the inhibitor (Am^{-2}).

5.2.2.2 Electrochemical Impedance Spectroscopy

Generally, EIS is used to investigate the amount of current flow and resistance value occurring on the metal surface with the presence and absence of inhibitors. EIS experiments are particularly suitable for rapidly and accurately obtaining corrosion rates [32–34]. The impedance of a system (frequency of a variable potential) is determined from EIS [35, 36]. Charge transfer resistance (Rct), solution resistance (Rs), and double-layer capacitance (Cdl) constitute the fundamental electrical equivalent circuit for characterizing the electrochemical performance of a substrate/metal in a corrosive solution. The EIS evaluation generated a Nyquist plot, with the imaginary part on the Y-axis and the real impedance component on the X-axis [37, 38]. The inhibition efficiency from EIS measurement can be calculated as shown in equation (5.5):

$$Inhibition\ efficiency = \frac{Ra - Rb}{Ra} x100 \tag{5.5}$$

where Ra is charge Rct in the absence of the inhibitor, and Rb is charge Rct in the presence of the inhibitor.

Shahmoradi et al. [39] are among the researchers working on corrosion inhibition investigation using the EIS method. They have studied the potential of pistachio nut extract as a corrosion inhibitor for mild steel in an HCl solution. In another study, Pal and Das [40] have elucidated that addition of banana flower bract extract for steel in an HCl and H_2SO_4 solutions can be a suitable corrosion inhibitor. The investigations revealed the maximum inhibition efficiencies of banana bract extract was 90.84% for 1 M HCl and 90.61% for 0.5 M H_2SO_4 media with 200 mg L^{-1} concentration of banana bract extract. In another study, Thomas et al. [41] reported the efficiency of *Garcinia* fruit rind extract as inhibitor for mild steel in HCl solution. As a comparison, the radius of the capacitive loop decreases with a higher concentration of HCl. This shows that the efficiency of *Garcinia* fruit rind is decreased due to the decrease of the energy barrier for corrosion. The higher concentration of HCl would also cause faster formation of the activated complex and lead to a higher corrosion rate. Additionally, the impedance decreases with temperature, elucidating that the rate of corrosion is increasing. The increment in temperature would increase the desorption rate of inhibitor molecules from the protective coating layer, leading to severe corrosion of the mild steel. As earlier highlighted, various methods have been used to determine the inhibition efficiency of agricultural waste extracts, such as WLM, PDP, and EIS techniques. The results exhibit good agreement between the values of corrosion efficiency as obtained from the listed methods. The outstanding performance of corrosion inhibitors from agricultural waste extracts is recognized by high corrosion inhibition efficiency, where the percentage efficiency results are mostly higher than 85%.

5.2.3 Surface Analysis in Corrosion

Scanning electron microscopy, atomic force microscope (AFM), X-ray photoelectron spectroscopy (XPS), Fourier transform infrared (FTIR), and energy dispersive EDX are some of the most common methods used for analyzing the surface characteristics of non-exposed and exposed substrate surfaces to the inhibitor. When corrosive media impacts an exposed surface without an inhibitor, the result is typically a rough surface. Conversely, the exposed metal surface with an inhibitor is smoother than the exposed metal surface without an inhibitor [42–44]. The presence of an inhibitor on a smooth metal surface is thought to be due to the production of an adsorptive and protective coating on it [45–48]. For instance, Shahmoradi et al. [49] have reported that the formation of an adsorptive film of walnut fruit husk extract on mild steel leads to inhibition in 1M HCl solution. A comparison between the surface of the metal with and without the addition of an inhibitor was reported in ref. [50]. The result supported the inhibition performance of the inhibitor on tin surface by SEM images. The corroded surface appears when the substrate is exposed to saline medium without an inhibitor. The author stated that the inhibitor extract can generate protective film and thus inhibits corrosion. SEM micrographs and the corresponding energy-dispersive spectroscopy (EDX) spectra would also show the surface characterizations with the chemical determination at the selected corroded surface. For instance, Bidi et al. [51] found the metal substrate's damaged surface without leech extract as an inhibitor. EDX was used to prove that the corroded surface is covered by corrosion products compared to the surface with leech extract. In another study, Sanni et al. [52] have further elucidated the formation of the protective layer on the metal surface. They found the EDX spectra containing different protective peaks with palm kernel shell extract. This finding supported the formation of metal oxides as corrosion products on the stainless steel surface with the absence of palm kernel shell extract. In addition, they have also proven the reduction of O and Cl peaks intensity with the presence of palm kernel shell extract. This further confirmed the formation of adsorbed inhibitor film onto the steel surface. Apart from SEM and EDX, an AFM could be used for three-dimensional (3D) quantitative morphological analysis of the inhibited and uninhibited specimens. Prasad et al. [53], Abdulridha et al. [54], and Ituen et al. [55] have successfully employed the AFM to measure the surface roughness of the steel specimens. They found that uninhibited and corroded metal specimens have average surface roughness between 336 and 380 nm. Upon inhibition process, the surface roughness of inhibited steel specimens could be reduced to as low as 25 nm, depending on the concentration of inhibitor used. Therefore, they have concluded that a suitable corrosion inhibitor would yield a lower surface roughness of the inhibited metal specimens. In addition, XPS is also a high-potential surface analysis to investigate the inhibitor-metal corrosion performance.

The XPS spectra analysis can identify elemental species in different oxidation states and with different chemical environments on the surface using a nonlinear least square curve-fitting procedure [56]. For example, Liu et al. [57] have investigated the effects of waste *Platanus acerifolia* leaves on steel substrate. They have successfully improved the corrosion-resistant *P. acerifolia* leaves extract performance on steel mainly by forming a carbonaceous organic film, as proven in XPS spectra analysis. The XPS analysis has also successfully revealed the presence of different oxidation states of iron on the steel sample. Therefore, they have concluded that the steel bar's surface is oxidized and a passivation film is formed on the surface. The iron oxide passivation film functioned as a corrosion protective layer on the steel surface.

Surface analysis techniques such as Auger electron spectroscopy (AES) and XPS can offer details on the atomic layers' outermost layers [58]. XPS may also identify the binding energy of components on the surface, which can provide insight into their chemical state. These methods are beneficial for studying thin protective scales, multilayer deposits, and subsurface diffusion into the base metal. Depth profiling is a valuable XPS technique, although ion mixing from the sputtering process can make interpretation difficult. Organic chemicals in corrosion products are frequently identified using FTIR spectroscopy and Raman spectroscopy. Both approaches can analyze microvolumes. Raman spectroscopy is a light scattering technique in which the molecule's chemical structure scatters the incident high-intensity laser light. As the analysis does not involve a vacuum, it can be beneficial for identifying some surface species. For example, Raman can be used to establish the type of coke and how it was created, which might be crucial for the continuous operation of a refinery or petrochemical facility [59]. Raman can also distinguish between distinct surface iron oxides, hydroxides and oxyhydroxides, aluminum oxides, copper corrosion products, and zirconium oxides on a metal. However, not all metal oxides or sulfides exhibit the required molecular vibrations or excitations for detection.

5.2.4 Theoretical Analysis in Corrosion

Based on the explanation in the previous section, most of the methods used for corrosion inhibition are time-consuming and expensive. The current computer technology is beneficial for a researcher to employ computer analysis, such as molecular simulations, to design or predict novel corrosion inhibitor performances [60–62]. Quantum chemical (QC) methods have proven to be highly useful in determining the molecular structure and elucidating potent inhibitors' electronic structure and reactivity [63, 64]. Besides effectively forecasting corrosion inhibition at the molecular level, QC calculations can reduce the research cost due to the selected blind screening test and raw materials [65]. In QC methods, density functional theory (DFT) and molecular dynamics (MD) simulation are usually employed to offer inhibition mechanisms at molecular-level insights. In DFT simulation, the highest occupied molecular orbital (HOMO) and lowest unoccupied molecular orbitals (LOMO) play a pivotal role in determining the efficacy of corrosion inhibition [66]. The energy of HOMO (E_{HOMO}), energy of LUMO (E_{LUMO}), gap energy (ΔE_{gap}), absolute chemical hardness (η), chemical softness (s), and altitude of the transferred electron (DN_{110}) can be determined using Gaussian 03W as shown in equation (5.6) [67].

$$\Delta E_{gap} = E_{LUMO} - E_{HOMO} \tag{5.6}$$

$$\eta = \frac{\Delta E_{gap}}{2} = \frac{E_{LUMO} - E_{HOMO}}{2} \tag{5.7}$$

$$\sigma = \frac{1}{\eta} = \frac{2}{E_{LUMO} - E_{HOMO}} \tag{5.8}$$

$$\Delta N_{110} = \frac{\varphi - X_{inh}}{2\left(\eta_{Fe} + \eta_{inh}\right)} \tag{5.9}$$

where φ, η_{Fe}, η_{inh}, and X_{inh} are the work function of metal surface, absolute hardness of iron, absolute hardness of inhibitor molecule, and absolute electronegativity of the inhibitor molecule, respectively.

For example, Saxena and Kumar [68] have utilized Hyperchem 8.0 software to draw the HOMO and LUMO orbitals of the *Sonchus oleraceus* extract and determine the energy of these orbitals. Their findings have proven that stronger inhibitor adsorption would yield a more significant energy gap between the HOMO and LUMO.

HOMO energy is related to the inhibitor molecules with the electron-donating ability, and the reduced LUMO value is associated with the electron acceptance from the higher energy level of metal d orbital. In other words, a larger ΔE_{gap} indicates high-inhibition performance and reflects the higher effectiveness of corrosion protection [69]. In another study, Shahmoradi et al. [70] have also applied similar methods to support the corrosion inhibitor performances of lignin and cellulose compounds extracted from agricultural waste on mild steel. They found that the increment N value could be related to many O heteroatoms and the phenyl rings containing lone pair and pi electrons in cellulose or lignin. Theoretically, the calculated DN values are higher than zero. It implies that the inhibitor molecules could transfer their electrons to the metallic surface, and the inhibition efficiency increases when the value of the shared electron is lower than 3.6. Besides that, Sanaei et al. [71] have also conducted MD simulations to extract molecular-level views concerning the inhibitor and metal substrate intermolecular interactions and the adsorption tendency of corrosion inhibitors. In visual inspection of the MD results, the inhibitor binding to the surface is also quantitatively studied by computing the interaction and binding energies. This determination will be able to detect the activity of corrosion inhibitors. The data from DFT and MD simulations provide further insight into the interaction mechanism.

5.2.4.1 Computational Chemistry Techniques

A range of chemical, electrochemical, and surface morphological approaches are commonly used to evaluate corrosion inhibitor inhibition efficiency. These experimental procedures have several flaws. These practical approaches are generally linked to the costly manufacture and consumption of harmful substances that harm the environment. These chemicals are generated in multicomponent reactions that require many solvents through the filtration and work-up phases from poisonous and expensive starting materials and catalysts. Excessive environmental discharge of organic solvents and toxic catalysts generates major environmental issues [72, 73]. Corrosion inhibitors are linked to producing various unwanted by-products that might harm the environment because of their sequential presence. Instead of yielding positive findings, most substances fail in the experimental trials due to their hazardous and expensive manufacture. Consequently, computational chemistry tools have recently been available for assessing the efficiency of corrosion inhibitors. The ecologically beneficial performance of such approaches is based on the concept that the inhibitory efficacy of chemicals may be forecast before their hazardous and costly manufacturing. In contrast to experimental methods, computational modeling uses software to predict inhibitory efficiency rather than expensive instrumentation [74–76].

Several computational chemistry approaches, particularly MD and DFT, have recently emerged as strong corrosion monitoring techniques. DFT simulations are the most advanced computing tool for determining essential indices such as energy of frontier molecular orbitals (FMOs; EHOMO and ELUMO) and numerous correlation parameters such as energy difference (ELUMO–EHOMO=E), global hardness, fraction of electron exchange (E), dipole moment, electronegativity, chemical softness and so on, which determines adsorption capacity and corrosion inhibition effectiveness. The DFT method is commonly used to determine the comparative inhibitory impact of a group of compounds with comparable chemical molecules. High inhibitory efficiency is typically correlated with greater EHOMO, dipole moment, chemical softness values, and ELUMO, E, electronegativity, and hardness values. E (ELUMO–EHOMO) is the most important indices in the energy band gap, and its smaller value is related to more inhibition efficacy. The EHOMO is commonly associated with the e-donating capability of the inhibitor molecule. Conversely, ELUMO is associated with the e-acceptance ability from the vacant d-orbital of the metallic outer layer. The energy difference (E) represents the inhibitor's chemical reactivity, with a lower value indicating more efficiency.

5.2.4.2 Ultrasonics

It is crucial to mention that some portions of the inhibitor molecule connect with the metallic surface during metal–inhibitor interactions. The application of DFT is recognized as the most vital method in discovering the binding sites of the inhibitor compound among the various experimental and computational methods [77, 78]. This approach involves applying a low-frequency sinusoidal potential to a three-electrode measuring device and evaluating the current intensity that occurs. Because the corrosion rate is exponential, a potential disruption by one or even more sine waves results in higher reaction frequencies than the applied signal. The current responses could be monitored at zero, fundamental, harmonic, and intermodular frequencies. The Faraday rectification (FR) technique is used to determine the DC frequency "zero" that can be utilized to compute the corrosion rate if any of the Tafel variables is specified. By analyzing the harmonic frequencies, a single test can determine the speed of corrosion and Tafel parameters. In Harmonic Distortion Analysis (HDA), the corrosion contact is subjected to a single, low-frequency, low-distortion voltage source. Three separate frequencies are employed as a quality check to ensure the approach is repeatable.

The amplitude ranges from 10 to 30 mV from peak to peak, and the most common frequency is 0.1–10 Hz. Ultrasonic testing has traditionally been used to determine thickness. A transducer is designed to vibrate at a higher frequency and is connected to any part of a sample surface whose width and thickness must be measured, either directly or indirectly. Combining electronics and computers greatly enhances ultrasonic equipment. Rugged devices using microcomputers and motor-driven robotic systems can measure the thickness of corroded item walls at thousands of places over $0.1m^2$ ($1\ ft^2$). Combined with the increased precision of site investigations permitted by computer algorithms, this feature has shown that the automated method is appropriate for online corrosion assessment. The ultrasonic method is primarily used during vessel soundness assessments. Moreover, it can be utilized as a corrosion monitor in circumstances of severe metal loss. With some significant advances in transducers, continuous monitoring is now possible. Although the approach can be utilized online, its sensitivity prevents it from being used for real-time observations. The magnetic flux leakage (MFL) technique employs a blend of permanent magnets and sensor coils to detect corrosion in steel plates, tubes, and pipes. As the flux conducts a circuit between the opposite poles of two magnets, it is funneled into the sample component, getting examined. When a metal defect arises, flux escapes and is detected by sensors mounted at the metal's surface and magnets. The sensors detect the amplitude of the leaking flux. The leaking flux amplitude correlates to recorded anomalies in related materials after inspection and wall-loss estimations are generated. Tanks, heat exchangers, pipelines, and other tubular components have all been examined using MFL. For various applications, different MFL systems are employed. Instruments for in-line pipeline evaluation are complex, self-contained devices that collect and store data from a range of meters and inches. MFL is a technique for tracking corrosion over time.

The results of ongoing inspections are collated, and the corrosion rate is measured. The MFL approach could locate faults involving volumetric metal loss, such as pitting. Calibration against the real part or component is required for this procedure. Radiographic imaging can determine the thickness of damaged pipes and other systems. The pit depth is measured by correlating the variation in the optical density of the coating in a non-corroded part and the pitted area. The variation in the corrosion thickness and area can be evaluated, and corrosion rates can be calculated by repeating surveys of selected areas at a frequency defined by the severity of corrosion. The method has been tested in extreme oil field conditions. As the approach does not require access to the component being inspected, it can be used to examine insulated, clad, bundled, or otherwise inaccessible pipework.

5.2.4.3 Electrical Field Mapping

Electrical field mapping can be used to detect pipe wall corrosion instantaneously. It includes detecting minor voltage fluctuations created by an induced current applied to a pipe to track changes in the defined region caused by surface cracking, corrosion, pitting, or erosion. The induced current is applied into the pipe 3–10m (about 9–30 ft) apart for broad pipes or a few millimeters or inches apart for small pipes to produce uniform current distribution. Electrical connections are made externally using contact pins that are welded, bonded, or spring-loaded. The voltage readings are contrasted to observe whether any nonuniformity could be caused by pitting or cracking in the assessed section. The average voltage drop is compared to a reference voltage to determine general metal loss. The pin spacing determines the monitored zone. The resolution for localized corrosion decreases as pin spacing increases. However, the resolution for broad corrosion increases. The thickness of the wall affects the resolution of general corrosion. On the web, the method is utilized to provide information regarding metal loss. The approach is helpful in inaccessible places, because no access is required after initial installation. The technique has no consumable parts except the pipe spool directly if used with high corrosion rates.

5.3 CHEMICAL ANALYSIS

Apart from theoretical calculation methods, chemical analyses such as infrared and ultraviolet (UV)-visible spectroscopic measurements were performed to investigate the chemical properties of inhibitors [79, 80]. For instance, complexes formed by the metal ions and functional groups in inhibitor participate in the corrosion inhibitory action. These complexes exhibit a restrictive effect on the side chains and, consequently, on the vibration modes of the CH_2 groups. UV–visible spectroscopic measurements were undertaken as it has been reported that a change in the position of the absorbance maximum and a shift in the absorbance value indicates that a complex between two species in solution is probably formed. This can be used as evidence of new product formation after the immersion of the substrate in a corrosive medium with an inhibitor. A comparison between the individual spectrum of corrosive medium with substrate, inhibitor, and a combination of substrate and inhibitor individually can differentiate the intensity and new band appearance. As reported by Palaniappan et al. [81], they have proven that UV–visible absorption spectroscopy supports the adsorption of extract molecules on the mild steel surface. Mobin et al. [82] investigated the presence of functional groups originating from *Cissus quadrangularis* plant extract and zirconium acetate in mild steel corrosion inhibition process in 1 M HCl solution using FTIR analysis. After a weight loss experiment, they compared the respective inhibitors' spectrum adsorbed on the steel surface, where the adsorbed inhibitors were scrapped off and subjected to FTIR analysis. They have also observed that significant peaks such as OH group of alcohol, OH stretch of carboxylic acid, carbonyl stretching, CN stretch amine, and CO stretch of ester are shifted toward lower wave number. These observations confirmed the interactions of constituent's presence in CQ with the metal.

5.4 BULK ANALYSIS IN CORROSION

Although EDX examination can offer local elemental analyses, it can also provide a bulk quantitative scale analysis. Such data can be used to identify and quantify trace elements and the potential metal source for the deposit if it is no longer adherent to the metal. X-ray fluorescence (XRF), inductively coupled plasma mass spectroscopy (ICP-MS), and atomic adsorption spectroscopy are examples of potential instrumental techniques. Various chromatographic instruments can be used to identify and measure ion components such as Cl^-, F^-, CO_3=, and NH_4^+ in corrosion deposits. These procedures necessitate specialized sample preparation. EDX analysis offers elemental information such as Fe and O but cannot distinguish between hematite and magnetite corrosion products. X-ray powder diffraction can be used to perform such identification. The angles and intensities of X-rays diffracted from the planes of the crystalline structure are distinctive with this approach, allowing identification of the various compounds in the corrosion scale. Because atomic substitution produces a modest lattice strain that can be difficult to detect, elemental analysis by XRD is highly beneficial for identifying chemicals present. As an example, Ca^{+2} ion substitution into $FeCO_3$ may be undetectable. As the morphology of corrosion products/scale crystals is platy, fibrous, or tabular, they may not randomly align during sample preparation, causing some planes to yield a more excellent intensity ratio than others and influencing quantification from peak intensity ratios. Each part must be quantified using a Rietveld refinement, which uses least squares analysis to refine a theoretical pattern until it fits the measured pattern. When the sample size is modest, distributing the corrosion products over a glass slide can frequently result in an acceptable diffraction pattern. Micro-focused XRD combined with SEM can enable elemental and phase identification of the same region. This method is particularly beneficial for multicolored scales and spatial distribution investigations.

5.5 CHALLENGE AND OUTLOOK

Numerous scientific findings on corrosion inhibitors have been published in the literature. Nevertheless, many related matters need to be considered; for instance, the standard of results regarding the performance of agricultural waste extracts as corrosion inhibitors. For example, statistical analysis is necessary to compare similar extracts as corrosion inhibitors in different published works. Besides, error bars or standard deviation should be calculated and displayed on graphs. This indicates that the reproducibility test is compulsory, and errors are recognized during the analysis. In addition, computational modeling and simulation could be employed to predict the performance of agricultural waste extracts as effective corrosion inhibitors. For instance, Monte Carlo, MD, and discreet Fourier transformed-based simulations and computational studies should be employed to examine the inhibitor-metal interactions thoroughly. Based on the literature, these computational methods are less reported than the experimental analyses. This theoretical method is required to support the experimental values obtained. Hence, both methods can be used as a strong relation of corrosion inhibitor from agricultural waste extract with more details. Lastly, future research should focus on the commercialization potential of corrosion inhibitors. It is generally agreed that larger-scale corrosion inhibitors' industrial production would require further in-depth tuning of formulation and processing parameters beyond the parameters discussed in the review article. The performances of commercial corrosion inhibitors, including their inhibition efficiency, must be thoroughly investigated and verified before application for commercial corrosion protection purposes. The research should include broader production, marketing, distribution, sales, and customer service processes. Analyzing and evaluating potential corrosion inhibitors should also be accompanied by financial-related and diligent analysis to predict and assess the impact and future of corrosion inhibitors in the economy sector, especially in the long run. These concerns are essential for attracting investors and other potential organizations to successfully commercialize and utilize the newly invented corrosion inhibitors in the global corrosion engineering market.

5.6 SUMMARY

Corrosion in metal and alloys is known to be a natural process, but it is hazardous to humans and the environment. The most practical way to tackle this problem is by using a corrosion inhibitor as a preventive means to slow down the corrosion process. The utilization of agricultural waste extracts as green corrosion inhibitors has been extensively reported in the literature as an alternative to conventional corrosion inhibitors. Numerous experimental and quantitative analytical methodologies have confirmed the superior efficiency of agricultural waste extracts as corrosion inhibitors owing to their structural analysis and molecule–surface interactions at an atomic level. Different experimental and theoretical analysis methods have verified the high performance of agricultural waste extracts as green corrosion inhibitors based on structural analysis and molecule–surface interactions at an atomic level. However, several issues need to be tackled to create full green production of green corrosion inhibitors, and the utilization or production of any toxic solution needs to be minimized. Lastly, the compilation and factors that influence the properties and performance of green corrosion inhibitors are aimed to help other researchers gain a broad insight into the importance of the evaluation of green corrosion inhibitors in preventing metal corrosion.

REFERENCES

1. Aghaaminiha, M., Mehrani, R., Colahan, M., Brown, B., Singer, M., Nesic, S., Vargas, S.M. and Sharma, S., 2021. Machine learning modeling of time-dependent corrosion rates of carbon steel in presence of corrosion inhibitors. *Corrosion Science*, *193*, 109904.
2. Ai, D., Du, L., Li, H. and Zhu, H., 2022. Corrosion damage identification for reinforced concrete beam using embedded piezoelectric transducer: Numerical simulation. *Measurement*, *192*, 110925.
3. Thakur, A., Sharma, S., Ganjoo, R., Assad, H. and Kumar, A., 2022, May. Anti-corrosive potential of the sustainable corrosion inhibitors based on biomass waste: A review on preceding and perspective research. *Journal of Physics: Conference Series*, *2267*(1), 012079.
4. Ranasinghe, K. and DeSilva, S., 2022. Combined use of non-destructive tests, visual inspection and service life prediction to ensure the reliability of reinforced concrete water tanks. *Journal of the National Science Foundation of Sri Lanka*, *50*(2), pp.371–386.
5. Singh, A., Ansari, K.R., Chauhan, D.S., Quraishi, M.A. and Kaya, S., 2020. Anti-corrosion investigation of pyrimidine derivatives as green and sustainable corrosion inhibitor for N80 steel in highly corrosive environment: Experimental and AFM/XPS study. *Sustainable Chemistry and Pharmacy*, *16*, 100257.
6. Molhi, A., Hsissou, R., Damej, M., Berisha, A., Thaçi, V., Belafhaili, A., Benmessaoud, M., Labjar, N. and El Hajjaji, S., 2021. Contribution to the corrosion inhibition of C38 steel in 1 M hydrochloric acid medium by a new epoxy resin PGEPPP. *International Journal of Corrosion and Scale Inhibition*, *10*(1), pp.399–418.
7. Bashir, S., Thakur, A., Lgaz, H., Chung, I.M. and Kumar, A., 2020. Corrosion inhibition efficiency of bronopol on aluminium in 0.5 M HCl solution: Insights from experimental and quantum chemical studies. *Surfaces and Interfaces*, *20*, 100542.
8. Hameed, R.S.A., Aljohani, M.M., Essa, A.B., Khaled, A., Nassar, A.M., Badr, M.M., Al-Mhyawi, S.R. and Soliman, M.S., 2021. Electrochemical techniques for evaluation of expired megavit drugs as corrosion inhibitor for steel in hydrochloric acid. *International Journal of Electrochemical Science*, *16*(4), 210446.
9. Danaee, I., RameshKumar, S., RashvandAvei, M. and Vijayan, M., 2020. Electrochemical and quantum chemical studies on corrosion inhibition performance of 2, 2'-(2-hydroxyethylimino) bis [N-(alphaalpha-dimethylphenethyl)-N-methylacetamide] on mild steel corrosion in 1M HCl solution. *Materials Research*, *23*.
10. Ali, I.H., 2021. Experimental, DFT and MD assessments of bark extract of *Tamarix aphylla* as corrosion inhibitor for carbon steel used in desalination plants. *Molecules*, *26*(12), p.3679.

11. Zhao, J., Tu, Z. and Chan, S.H., 2021. Carbon corrosion mechanism and mitigation strategies in a proton exchange membrane fuel cell (PEMFC): A review. *Journal of Power Sources*, *488*, 229434.
12. Zhao, F., Cui, C., Dong, S., Xu, X. and Liu, H., 2023. An overview on the corrosion mechanisms and inhibition techniques for amine-based post-combustion carbon capture process. *Separation and Purification Technology*, *304*, 122091.
13. Li, Q., Zhang, Y., Cheng, Y., Zuo, X., Wang, Y., Yuan, X. and Huang, H., 2022. Effect of temperature on the corrosion behavior and corrosion resistance of copper–aluminum laminated composite plate. *Materials*, *15*(4), p.1621.
14. McCormack, M.A., McFee, W.E., Whitehead, H.R., Piwetz, S. and Dutton, J., 2022. Exploring the use of SEM–EDS analysis to measure the distribution of major, minor, and trace elements in bottlenose dolphin (*Tursiops truncatus*) teeth. *Biological Trace Element Research*, *200*, pp.1–13.
15. Kleiner, F., Decker, M., Rößler, C., Hilbig, H. and Ludwig, H.M., 2022. Combined LA-ICP-MS and SEM-EDX analyses for spatially resolved major, minor and trace element detection in cement clinker phases. *Cement and Concrete Research*, *159*, 106875.
16. Sharma, S. and Kumar, A., 2021. Recent advances in metallic corrosion inhibition: A review. *Journal of Molecular Liquids*, *322*, 114862.
17. Vu, N.S.H., Binh, P.M.Q., Dao, V.A., Thu, V.T.H., Van Hien, P., Panaitescu, C. and Nam, N.D., 2020. Combined experimental and computational studies on corrosion inhibition of *Houttuynia cordata* leaf extract for steel in HCl medium. *Journal of Molecular Liquids*, *315*, 113787.
18. Salleh, S.Z., Yusoff, A.H., Zakaria, S.K., Taib, M.A.A., Seman, A.A., Masri, M.N., Mohamad, M., Mamat, S., Sobri, S.A., Ali, A. and Ter Teo, P., 2021. Plant extracts as green corrosion inhibitor for ferrous metal alloys: A review. *Journal of Cleaner Production*, *304*, 127030.
19. Asfia, M.P., Rezaei, M. and Bahlakeh, G., 2020. Corrosion prevention of AISI 304 stainless steel in hydrochloric acid medium using garlic extract as a green corrosion inhibitor: Electrochemical and theoretical studies. *Journal of Molecular Liquids*, *315*, 113679.
20. Naseef Jasim, A., Abdulhussein, B.A., Mohammed Noori Ahmed, S., Al-Azzawi, W.K., Hanoon, M.M., Abbass, M.K. and Al-Amiery, A.A., 2023. Schiff's base performance in preventing corrosion on mild steel in acidic conditions. *Progress in Color, Colorants and Coatings*, *16*(4), pp.319–329.
21. de Oliveira, G.A., Teixeira, V.M., da Cunha, J.N., dos Santos, M.R., Paiva, V.M., Rezende, M.J.C., do Valle, A.F. and D'Elia, E., 2021. Biomass of microalgae *Chlorella sorokiniana* as green corrosion inhibitor for mild steel in HCl solution. *International Journal of Electrochemical Science*, *16*(2), 210249.
22. Sanni, O., Popoola, A.P.I. and Fayomi, O.S.I., 2021. Adsorption ability and corrosion inhibition mechanism of agricultural waste on stainless steel in chloride contaminated environment. *Materials Today: Proceedings*, *43*, pp.2215–2221.
23. Yaqo, E.A., Anaee, R.A., Abdulmajeed, M.H., Tomi, I.H.R. and Kadhim, M.M., 2020. Potentiodynamic polarization, surface analyses and computational studies of a 1, 3, 4-thiadiazole compound as a corrosion inhibitor for Iraqi kerosene tanks. *Journal of Molecular Structure*, *1202*, 127356.
24. Pragathiswaran, C., Ramadevi, P. and Kumar, K.K., 2021. Imidazole and Al3+ nano material as corrosion inhibitor for mild steel in hydrochloric acid solutions. *Materials Today: Proceedings*, *37*, pp.2912–2916.

25. Odusote, J.K., Asafa, T.B., Oseni, J.G., Adeleke, A.A., Adediran, A.A., Yahya, R.A., Abdul, J.M. and Adedayo, S.A., 2021. Inhibition efficiency of gold nanoparticles on corrosion of mild steel, stainless steel and aluminium in 1M HCl solution. *Materials Today: Proceedings*, *38*, pp.578–583.
26. Anor, O., Lahmady, S., Forsal, I., Hanine, H., Ourradi, H. and Elharami, A., 2022. An experimental investigation of a date seeds hydro-acetonic mixture extract inhibitor for corrosion inhibition of carbon steel in an acidic medium at high temperatures. *Biointerface Research in Applied Chemistry*, *13*, pp.1–13.
27. Şahin, E.A., Solmaz, R., Gecibesler, I.H. and Kardaş, G., 2020. Adsorption ability, stability and corrosion inhibition mechanism of phoenix dactylifera extract on mild steel. *Materials Research Express*, *7*(1), 016585.
28. Pahuja, P., Saini, N., Chaouiki, A., Salghi, R., Kumar, S. and Lata S., 2020. The protection mechanism offered by Heterophragma adenophyllum extract against Fe-C steel dissolution at low pH: Computational, statistical and electrochemical investigations. *Bioelectrochemistry*, *132*, 107400.
29. Badiea, A.M. and Mohana, K.N., 2009. Effect of temperature and fluid velocity on corrosion mechanism of low carbon steel in presence of 2-hydrazino-4, 7-dimethylbenzothiazole in industrial water medium. *Corrosion Science*, *51*(9), pp.2231–2241.
30. Sanni, O., Popoola, A.P.I. and Fayomi, O.S.I., 2018. Enhanced corrosion resistance of stainless steel type 316 in sulphuric acid solution using eco-friendly waste product. *Results in Physics*, *9*, pp.225–230.
31. Salleh, S.Z., Yusoff, A.H., Zakaria, S.K., Taib, M.A.A., Seman, A.A., Masri, M.N., Mohamad, M., Mamat, S., Sobri, S.A., Ali, A. and Ter Teo, P., 2021. Plant extracts as green corrosion inhibitor for ferrous metal alloys: A review. *Journal of Cleaner Production*, *304*, 127030.
32. Zhang, M., Liu, Y., Li, D., Cui, X., Wang, L., Li, L. and Wang, K., 2023. Electrochemical impedance spectroscopy: A new chapter in the fast and accurate estimation of the state of health for lithium-ion batteries. *Energies*, *16*(4), p.1599.
33. Feliu Jr, S., 2020. Electrochemical impedance spectroscopy for the measurement of the corrosion rate of magnesium alloys: Brief review and challenges. *Metals*, *10*(6), p.775.
34. Alvarez, P.E., Fiori-Bimbi, M.V., Neske, A., Brandán, S.A. and Gervasi, C.A., 2018. Rollinia occidentalis extract as green corrosion inhibitor for carbon steel in HCl solution. *Journal of Industrial and Engineering Chemistry*, *58*, pp.92–99.
35. Choi, W., Shin, H.C., Kim, J.M., Choi, J.Y. and Yoon, W.S., 2020. Modeling and applications of electrochemical impedance spectroscopy (EIS) for lithium-ion batteries. *Journal of Electrochemical Science and Technology*, *11*(1), pp.1–13.
36. Liu, C., Bi, Q., Leyland, A. and Matthews, A., 2003. An electrochemical impedance spectroscopy study of the corrosion behaviour of PVD coated steels in 0.5 N NaCl aqueous solution: Part II.: EIS interpretation of corrosion behaviour. *Corrosion Science*, *45*(6), pp.1257–1273.
37. Magar, H.S., Hassan, R.Y. and Mulchandani, A., 2021. Electrochemical impedance spectroscopy (EIS): Principles, construction, and biosensing applications. *Sensors*, *21*(19), p.6578.
38. Laschuk, N.O., Easton, E.B. and Zenkina, O.V., 2021. Reducing the resistance for the use of electrochemical impedance spectroscopy analysis in materials chemistry. *RSC Advances*, *11*(45), pp.27925–27936.
39. Shahmoradi, A.R., Talebibahmanbigloo, N., Javidparvar, A.A., Bahlakeh, G. and Ramezanzadeh, B., 2020. Studying the adsorption/inhibition impact of the cellulose

and lignin compounds extracted from agricultural waste on the mild steel corrosion in HCl solution. *Journal of Molecular Liquids*, *304*, 112751.

40. Pal, A. and Das, C., 2022. Investigations on corrosion inhibition in acidic media for BQ steel using banana flower bract, an eco-friendly novel agro-waste: Experimental and theoretical considerations. *Inorganic Chemistry Communications*, *145*, 110024.
41. Thomas, A., Prajila, M., Shainy, K.M. and Joseph, A., 2020. A green approach to corrosion inhibition of mild steel in hydrochloric acid using fruit rind extract of *Garcinia indica* (Binda). *Journal of Molecular Liquids*, *312*, 113369.
42. Murmu, M., Saha, S.K., Bhaumick, P., Murmu, N.C., Hirani, H. and Banerjee, P., 2020. Corrosion inhibition property of azomethine functionalized triazole derivatives in 1 mol L−1 HCl medium for mild steel: Experimental and theoretical exploration. *Journal of Molecular Liquids*, *313*, 113508.
43. Sedik, A., Lerari, D., Salci, A., Athmani, S., Bachari, K., Gecibesler, İ.H. and Solmaz, R., 2020. Dardagan Fruit extract as eco-friendly corrosion inhibitor for mild steel in 1 M HCl: Electrochemical and surface morphological studies. *Journal of the Taiwan Institute of Chemical Engineers*, *107*, pp.189–200.
44. Vu, N.S.H., Binh, P.M.Q., Dao, V.A., Thu, V.T.H., Van Hien, P., Panaitescu, C. and Nam, N.D., 2020. Combined experimental and computational studies on corrosion inhibition of *Houttuynia cordata* leaf extract for steel in HCl medium. *Journal of Molecular Liquids*, *315*, 113787.
45. Parthipan, P., Cheng, L. and Rajasekar, A., 2021. *Glycyrrhiza glabra* extract as an eco-friendly inhibitor for microbiologically influenced corrosion of API 5LX carbon steel in oil well produced water environments. *Journal of Molecular Liquids*, *333*, 115952.
46. Haruna, K., Alhems, L.M. and Saleh, T.A., 2021. Graphene oxide grafted with dopamine as an efficient corrosion inhibitor for oil well acidizing environments. *Surfaces and Interfaces*, *24*, 101046.
47. Caldona, E.B., Zhang, M., Liang, G., Hollis, T.K., Webster, C.E., Smith Jr, D.W. and Wipf, D.O., 2021. Corrosion inhibition of mild steel in acidic medium by simple azole-based aromatic compounds. *Journal of Electroanalytical Chemistry*, *880*, 114858.
48. Sanni, O., Iwarere, S.A. and Daramola, M.O., 2023. Investigation of eggshell agro-industrial waste as a potential corrosion inhibitor for mild steel in oil and gas industry. *Sustainability*, *15*(7), p.6155.
49. Shahmoradi, A.R., Ranjbarghanei, M., Javidparvar, A.A., Guo, L., Berdimurodov, E. and Ramezanzadeh, B., 2021. Theoretical and surface/electrochemical investigations of walnut fruit green husk extract as effective inhibitor for mild-steel corrosion in 1M HCl electrolyte. *Journal of Molecular Liquids*, *338*, 116550.
50. El Ibrahimi, B., Baddouh, A., Oukhrib, R., El Issami, S., Hafidi, Z. and Bazzi, L., 2021. Electrochemical and in silico investigations into the corrosion inhibition of cyclic amino acids on tin metal in the saline environment. *Surfaces and Interfaces*, *23*, 100966.
51. Bidi, M.A., Azadi, M. and Rassouli, M., 2021. An enhancement on corrosion resistance of low carbon steel by a novel bio-inhibitor (leech extract) in the H_2SO_4 solution. *Surfaces and Interfaces*, *24*, 101159.
52. Sanni, O., Ren, J. and Jen, T.C., 2022. Understanding the divergence of agro waste extracts on the microstructure, mechanical, and corrosion performance of steel alloy. *South African Journal of Chemical Engineering*, *40*(1), pp.57–69.
53. Prasad, D., Singh, R., Kaya, S. and El Ibrahimi, B., 2022. Natural corrosion inhibitor of renewable eco-waste for SS-410 in sulfuric acid medium: Adsorption, electrochemical, and computational studies. *Journal of Molecular Liquids*, *351*, 118671.

54. Abdulridha, A.A., Allah, M.A.A.H., Makki, S.Q., Sert, Y., Salman, H.E. and Balakit, A.A., 2020. Corrosion inhibition of carbon steel in 1 M H2SO4 using new Azo Schiff compound: Electrochemical, gravimetric, adsorption, surface and DFT studies. *Journal of Molecular Liquids*, *315*, 113690.
55. Ituen, E., Yuanhua, L., Singh, A. and Li, R., 2021. Chemical modification of waste *Allium cepa* peels to Cu-complex composite and application as eco environmental oilfield anticorrosion additive. *Journal of King Saud University-Engineering Sciences*, *33*(6), pp.375–385.
56. Cabello Mendez, J.A., Pérez Bueno, J.D.J., Meas Vong, Y. and Portales Martínez, B., 2022. Cerium compounds coating as a single self-healing layer for corrosion inhibition on aluminum 3003. *Sustainability*, *14*(22), 15056.
57. Liu, Q., Song, Z., Han, H., Donkor, S., Jiang, L., Wang, W. and Chu, H., 2020. A novel green reinforcement corrosion inhibitor extracted from waste *Platanus acerifolia* leaves. *Construction and Building Materials*, *260*, 119695.
58. Wu, X., Wiame, F., Maurice, V. and Marcus, P., 2020. 2-Mercaptobenzimidazole films formed at ultra-low pressure on copper: Adsorption, thermal stability and corrosion inhibition performance. *Applied Surface Science*, *527*, 146814.
59. Moya-Cancino, J.G., Honkanen, A.P., Van Der Eerden, A.M., Oord, R., Monai, M., Ten Have, I., Sahle, C.J., Meirer, F., Weckhuysen, B.M., De Groot, F.M. and Huotari, S., 2021. In Situ X-ray Raman scattering spectroscopy of the formation of cobalt carbides in a Co/TiO_2 Fischer–Tropsch synthesis catalyst. *ACS Catalysis*, *11*(2), pp.809–819.
60. Ebenso, E.E., Verma, C., Olasunkanmi, L.O., Akpan, E.D., Verma, D.K., Lgaz, H., Guo, L., Kaya, S. and Quraishi, M.A., 2021. Molecular modelling of compounds used for corrosion inhibition studies: A review. *Physical Chemistry Chemical Physics*, *23*(36), pp.19987–20027.
61. Quadri, T.W., Olasunkanmi, L.O., Fayemi, O.E., Akpan, E.D., Verma, C., Sherif, E.S.M., Khaled, K.F. and Ebenso, E.E., 2021. Quantitative structure activity relationship and artificial neural network as vital tools in predicting coordination capabilities of organic compounds with metal surface: A review. *Coordination Chemistry Reviews*, *446*, 214101.
62. Baghel, D. and Banjare, M.K., 2023. Molecular modeling in corrosion inhibition assessment. In *Computational Modelling and Simulations for Designing of Corrosion Inhibitors* (pp. 79–92). Elsevier.
63. Grillo, I.B., Bachega, J.F.R., Timmers, L.F.S., Caceres, R.A., de Souza, O.N., Field, M.J. and Rocha, G.B., 2020. Theoretical characterization of the shikimate 5-dehydrogenase reaction from *Mycobacterium tuberculosis* by hybrid QC/MM simulations and quantum chemical descriptors. *Journal of Molecular Modeling*, *26*, pp.1–12.
64. Grillo, I.B., Urquiza-Carvalho, G.A. and Rocha, G.B., 2023. Quantum chemical descriptors as a modeling framework for large biological structures. In *Chemical Reactivity* (pp. 59–88). Elsevier.
65. Zhang, N., Zhang, J., Wang, G., Ning, X., Meng, F., Li, C., Ye, L. and Wang, C., 2022. Physicochemical characteristics of three-phase products of low-rank coal by hydrothermal carbonization: Experimental research and quantum chemical calculation. *Energy*, *261*, 125347.
66. Zhu, J., Zhou, G., Niu, F., Shi, Y., Du, Z., Lu, G. and Liu, Z., 2022. Understanding the inhibition performance of novel dibenzimidazole derivatives on Fe (110) surface: DFT and MD simulation insights. *Journal of Materials Research and Technology*, *17*, pp.211–222.

67. Hsissou, R., Benhiba, F., Khudhair, M., Berradi, M., Mahsoune, A., Oudda, H., El Harfi, A., Obot, I.B. and Zarrouk, A., 2020. Investigation and comparative study of the quantum molecular descriptors derived from the theoretical modeling and Monte Carlo simulation of two new macromolecular polyepoxide architectures TGEEBA and HGEMDA. *Journal of King Saud University-Science*, *32*(1), pp.667–676.
68. Saxena, A. and Kumar, J., 2020. Phytochemical screening, metal-binding studies and applications of floral extract of *Sonchus oleraceus* as a corrosion inhibitor. *Journal of Bio-and Tribo-Corrosion*, *6*, pp.1–10.
69. Hsissou, R., Benhiba, F., Dagdag, O., El Bouchti, M., Nouneh, K., Assouag, M., Briche, S., Zarrouk, A. and Elharfi, A., 2020. Development and potential performance of prepolymer in corrosion inhibition for carbon steel in 1.0 M HCl: Outlooks from experimental and computational investigations. *Journal of Colloid and Interface Science*, *574*, pp.43–60.
70. Shahmoradi, A.R., Talebibahmanbigloo, N., Javidparvar, A.A., Bahlakeh, G. and Ramezanzadeh, B., 2020. Studying the adsorption/inhibition impact of the cellulose and lignin compounds extracted from agricultural waste on the mild steel corrosion in HCl solution. *Journal of Molecular Liquids*, *304*, 112751.
71. Sanaei, Z., Ramezanzadeh, M., Bahlakeh, G. and Ramezanzadeh, B., 2019. Use of *Rosa canina* fruit extract as a green corrosion inhibitor for mild steel in 1 M HCl solution: A complementary experimental, molecular dynamics and quantum mechanics investigation. *Journal of Industrial and Engineering Chemistry*, *69*, pp.18–31.
72. Verma, C., Hussain, C.M., Quraishi, M.A. and Alfantazi, A., 2022. Green surfactants for corrosion control: Design, performance and applications. *Advances in Colloid and Interface Science*, *311*, 102822.
73. Quraishi, M.A., Chauhan, D.S. and Ansari, F.A., 2021. Development of environmentally benign corrosion inhibitors for organic acid environments for oil-gas industry. *Journal of Molecular Liquids*, *329*, 115514.
74. Adnouni, M., Jiang, L., Zhang, X.J., Zhang, L.Z., Pathare, P.B. and Roskilly, A.P., 2023. Computational modelling for decarbonised drying of agricultural products: Sustainable processes, energy efficiency, and quality improvement. *Journal of Food Engineering*, *338*, 111247.
75. Edache, E.I., Uzairu, A., Mamza, P.A. and Shallangwa, G.A., 2022. Computational modeling and analysis of the theoretical structure of thiazolino 2-pyridone amide inhibitors for *Yersinia pseudotuberculosis* and *Chlamydia trachomatis* infectivity. *Bulletin of Scientific Research*, *4*(1), pp.14–39.
76. Rani, A.J., Thomas, A., Williams, L. and Joseph, A., 2022. Effect of lunamarine, the major constituent of Boerhaavia diffusa leave extract on the corrosion inhibition of mild steel in hydrochloric acid; computational modelling, surface screening and electroanalytical studies. *Journal of Bio-and Tribo-Corrosion*, *8*, pp.1–21.
77. Lgaz, H., Saha, S.K., Chaouiki, A., Bhat, K.S., Salghi, R., Banerjee, P., Ali, I.H., Khan, M.I. and Chung, I.M., 2020. Exploring the potential role of pyrazoline derivatives in corrosion inhibition of mild steel in hydrochloric acid solution: Insights from experimental and computational studies. *Construction and Building Materials*, *233*, 117320.
78. Rasheed, S., Aziz, M., Saeed, A., Ejaz, S.A., Channar, P.A., Zargar, S., Abbas, Q., Alanazi, H., Hussain, M., Alharbi, M. and Kim, S.J., 2022. Analysis of 1-aroyl-3-[3-chloro-2-methylphenyl] thiourea hybrids as potent urease inhibitors: Synthesis, biochemical evaluation and computational approach. *International Journal of Molecular Sciences*, *23*(19), 11646.
79. Rbaa, M., Galai, M., Abousalem, A.S., Lakhrissi, B., Touhami, M.E., Warad, I. and Zarrouk, A., 2020. Synthetic, spectroscopic characterization, empirical and theoretical

investigations on the corrosion inhibition characteristics of mild steel in molar hydrochloric acid by three novel 8-hydroxyquinoline derivatives. *Ionics*, *26*, pp.503–522.
80. Laabaissi, T., Rbaa, M., Benhiba, F., Rouifi, Z., Kumar, U.P., Bentiss, F., Oudda, H., Lakhrissi, B., Warad, I. and Zarrouk, A., 2021. Insight into the corrosion inhibition of new benzodiazepine derivatives as highly efficient inhibitors for mild steel in 1 M HCl: Experimental and theoretical study. *Colloids and Surfaces A: Physicochemical and Engineering Aspects*, *629*, 127428.
81. Palaniappan, N., Cole, I., Caballero-Briones, F., Manickam, S., Thomas, K.J. and Santos, D., 2020. Experimental and DFT studies on the ultrasonic energy-assisted extraction of the phytochemicals of *Catharanthus roseus* as green corrosion inhibitors for mild steel in NaCl medium. *RSC Advances*, *10*(9), pp.5399–5411.
82. Mobin, M., Basik, M. and Shoeb, M., 2019. A novel organic-inorganic hybrid complex based on *Cissus quadrangularis* plant extract and zirconium acetate as a green inhibitor for mild steel in 1 M HCl solution. *Applied Surface Science*, *469*, pp.387–403.

6 Commercialization of Environmentally Sustainable Corrosion Inhibitors

6.1 INTRODUCTION: CORROSION AND ITS IMPACT

Corrosion is a natural process that occurs when materials, especially metals, degrade due to chemical reactions with their environment [1, 2]. Metallic materials are used in construction and industries like petroleum, oil, and gas. However, they are thermodynamically unstable and can easily undergo corrosive degradation due to environmental reactions. These reactions often involve moisture, oxygen, and contaminants, leading to the formation of corrosion products that weaken the structural integrity and functionality of materials. Corrosion can manifest in various forms, such as uniform corrosion, localized corrosion (such as pitting and crevice corrosion), galvanic corrosion, and stress corrosion cracking. The type and rate of corrosion depend on factors such as material composition, environmental conditions, temperature, and exposure to corrosive agents [3, 4].

Corrosion poses significant challenges to key industries, including the oil and gas industry, automotive industry, infrastructure, and manufacturing. Corrosion-related failures can result in leaks, spills, and costly downtime, jeopardizing safety and profitability. In the automotive sector, corrosion affects vehicle bodies, chassis, and critical components, leading to safety concerns, decreased resale value, and increased maintenance costs. Infrastructure, such as bridges, highways, and buildings, poses a substantial threat to public safety and economic stability, undermining the durability and longevity of these structures [5, 6]. During the 1960s, surface-active chelates, polyphosphates, polyphosphonates, phosphonates, and carboxylates were used as precipitating inhibitors [7, 8]. However, ecological considerations led to the replacement of toxic chemicals with natural alternatives like biopolymers, biosurfactants, vitamins, tannins, and natural compounds. Since 1995, environment-friendly approaches have focused on using rare earth metals (REMs), polyfunctional compounds, and encapsulation of inhibitors [9]. Organic compounds are now deemed effective and profitable methods of corrosion inhibition due to their efficiency, economy, ecology, and environmental friendliness. However, their limited solubility, especially in polar electrolytes, poses challenges.

In manufacturing processes, corrosion can damage machinery, equipment, and tools, leading to production disruptions and increased maintenance expenses [10]. To mitigate the widespread impact of corrosion, it is imperative to develop and implement

DOI: 10.1201/9781003441151-6

effective corrosion inhibition strategies. Corrosion inhibitors function by forming protective layers on the material's surface or altering the chemical reactions that drive corrosion [11, 12]. The significance of effective corrosion inhibitors is emphasized by several key factors: economic impact, safety, environmental concerns, and sustainability. Corrosion-induced structural failures can have catastrophic consequences, endangering lives and the environment. By preventing corrosion, inhibitors contribute to environmental protection and sustainability. In conclusion, corrosion is a pervasive and costly challenge that affects various industries, and research and development efforts aimed at advancing corrosion inhibition technologies are critical for the well-being and resilience of these industries.

6.2 GLOBAL CORROSION INHIBITOR MARKET

The global cost of corrosion is estimated to be around US$2.5 trillion, accounting for 3.5% of the world's gross domestic product (GDP). Direct costs include repair, storage, replacement, and modification of alloys. Indirect costs include leakage of liquids and gases from transport pipelines, affecting machinery performance and efficiency. The global corrosion inhibitor market is a vital segment of the chemical industry, driven by the need to combat corrosion-related issues across various sectors. The market has experienced substantial growth over the past decade due to the increasing awareness of corrosion's economic and safety implications. As of 2020, the global corrosion inhibitor market was valued at US$ 7.7 billion, reflecting a steady rise from previous years [13–15]. Key trends driving the growth of the corrosion inhibitor market include increased industrialization, infrastructure development, technological advancements, focus on sustainable solutions, and globalization. Emerging economies, such as oil and gas, petrochemicals, and manufacturing, require effective corrosion protection to safeguard their assets and maintain operational efficiency. Technological advancements have led to the creation of more advanced and environmentally friendly corrosion inhibitors, offering superior performance and reduced environmental impact.

Increasing environmental concerns have prompted industries to seek sustainable corrosion inhibition options, leading to the emergence of eco-friendly inhibitors, such as green corrosion inhibitors derived from natural sources [16–18]. Globalization has created a demand for corrosion inhibitors that can withstand diverse environmental conditions and corrosive agents, leading to a more competitive market. Major players in the market include BASF SE, AkzoNobel N.V., Ecolab Inc., Henkel AG & Co. KGaA, and Solvay S.A., each contributing significantly to market growth. Factors driving market expansion include industrial expansion, infrastructure investments, technological advancements, and stricter environmental regulations. As industries expand and modernize, the need for corrosion protection becomes increasingly critical to ensure asset longevity and reliability [19–20]. The global corrosion inhibitor market is experiencing significant growth due to increased industrialization, infrastructure development, technological advancements, and a shift toward sustainable solutions. Major players in the industry continue to innovate and expand their product offerings to meet the diverse needs of various sectors.

6.3 MARKETPLACE DYNAMICS OF CORROSION INHIBITORS

The corrosion inhibitor market is shaped by diverse factors, including technological advancements, regulatory requirements, and emerging markets [21]. Technological advancements have led to the development of more effective and environmentally friendly corrosion inhibition solutions, such as nanotechnology, smart coatings, green chemistry, corrosion monitoring, and computational modeling [22–24]. Regulatory requirements and standards significantly affect the corrosion inhibitor market, with stringent environmental regulations pushing the industry towards greener and less toxic corrosion inhibitors. Compliance with safety standards is nonnegotiable in industries where safety is paramount, such as oil and gas. Clear and accurate product labeling and documentation are essential for compliance and customer trust, and international harmonization facilitates international trade.

Emerging markets are witnessing growth in the corrosion inhibitor industry due to factors such as industrialization, infrastructure development, and increased awareness of corrosion-related challenges [25]. Key trends in emerging markets include the Asia-Pacific region, Middle East and North Africa (MENA), Latin America, and Africa. The rapid industrialization in countries like China and India has created a substantial demand for corrosion inhibition solutions, driven by infrastructure projects, manufacturing, and increased oil and gas exploration [13, 26]. The MENA region drives the demand for corrosion inhibitors in the oil and gas industry, with infrastructure development and petrochemical projects contributing to market growth [27]. Latin America's growing investments in infrastructure, particularly in countries like Brazil, have bolstered the corrosion inhibitor market, with transportation, construction, and energy sectors being prominent consumers of corrosion inhibitors [28].

The marketplace dynamics of corrosion inhibitors are influenced by technological advancements, regulatory requirements, and emerging markets. Technological innovations continue to drive the development of more effective and sustainable corrosion inhibition solutions. Regulatory compliance is essential for market access, and companies must adapt to evolving standards. Emerging markets present significant opportunities for growth as industries expand and prioritize corrosion prevention. Staying attuned to these dynamics is crucial for businesses operating in the corrosion inhibitor industry to remain competitive and capitalize on emerging trends.

6.4 CHALLENGES AND FUTURE OUTLOOK

This chapter evaluates the economic viability of sustainable corrosion inhibitors, considering factors such as production costs, scalability, and cost savings. It explores the current market landscape, potential barriers to market entry, and regulatory frameworks that may affect their adoption. It also identifies key stakeholders in the industry, assesses the competitive landscape, and compares the economic feasibility of sustainable and traditional corrosion inhibitors.

6.4.1 Drawbacks of Waste Extracts as Corrosion Inhibitors

Waste extracts from plants and industrial by-products have emerged as promising candidates for environmentally sustainable corrosion inhibitors due to their eco-friendliness and cost-effectiveness [29]. However, their utilization comes with several drawbacks that need careful consideration and mitigation.

1. Inconsistent composition: Waste extracts are complex mixtures of compounds, leading to inconsistent inhibition performance. Researchers are exploring methods to standardize the composition of waste extracts, such as identifying key active compounds responsible for corrosion inhibition [12]. This can be done in a more controlled manner to ensure consistency.
2. Limited effectiveness: Waste extracts may demonstrate corrosion inhibition properties but have lower effectiveness compared to synthetic inhibitors due to the presence of diverse chemical components [30]. Researchers are working on improving the effectiveness of waste extracts by optimizing formulations, identifying synergistic combinations of compounds, improving adsorption onto metal surfaces, and employing encapsulation techniques to enhance controlled release of active compounds.
3. Limited understanding of mechanisms: Waste extracts may lack a clear mechanistic understanding of their corrosion inhibition properties, making it challenging to optimize performance and predict behavior under different environmental conditions [31]. To overcome this limitation, researchers are conducting comprehensive studies to unravel the mechanisms of corrosion inhibition by waste extracts using advanced analytical techniques like spectroscopy and surface analysis.
4. Environmental impact: Waste extracts are considered environmentally friendly compared to many synthetic inhibitors, but their production and extraction processes can still have environmental implications. Researchers are exploring greener extraction methods, such as using supercritical fluids or reducing solvent usage, and optimizing the cultivation of plants for extraction.
5. Stability and shelf life: Many waste extracts are susceptible to degradation over time, reducing their shelf life and long-term effectiveness as corrosion inhibitors. Researchers are investigating methods to enhance the stability of waste extracts, such as encapsulation in protective matrices or the development of microencapsulation techniques, to extend the shelf life of waste extract-based corrosion inhibitors. Waste extracts from plants and industrial byproducts hold great promise as environmentally sustainable corrosion inhibitors [32]. However, their drawbacks, including inconsistent composition, limited effectiveness, lack of mechanistic understanding, environmental impact, and stability issues, must be addressed to fully harness their potential.

6.4.2 Agriculture, Biomass, Biopolymer Wastes, and Essential Oils as Green Corrosion Inhibitors: Challenges and Prospects

The utilization of agriculture, biomass, biopolymer wastes, and essential oils as green corrosion inhibitors presents a promising avenue for sustainable corrosion protection strategies [33, 34]. These natural resources are abundant and renewable, making them an attractive option for corrosion inhibition. However, challenges such as extraction complexity, variable composition, and formulation issues need to be addressed. Agricultural waste materials, such as rice husks, corn cobs, and sugarcane bagasse, contain natural compounds rich in lignin, cellulose, and hemicellulose, which can be harnessed for corrosion inhibition [35, 36]. However, extracting active corrosion inhibitors from agricultural waste materials can be complex and resource-intensive, and developing efficient extraction methods is a key challenge. The composition of agricultural waste can vary based on crop type, growth conditions, and postharvest treatment, making it difficult to predict inhibitor performance.

Biomass waste materials, such as wood chips and sawdust, contain natural compounds that can inhibit corrosion processes [37, 38]. However, they can show variability in composition, which can affect inhibitor performance. Extracting active corrosion inhibitors from biomass waste materials while maintaining their effectiveness can be challenging [39–41]. Formulating inhibitors from biomass waste that are compatible with various application methods, such as coatings or paints, is a key challenge.

Biopolymer waste materials, including chitosan and starch, have shown promise as corrosion inhibitors [42]. They are biodegradable and can form protective films on metal surfaces. However, achieving consistent and uniform film formation on metal surfaces can be challenging. Some biopolymers have limited solubility in water or common solvents, which can impact formulation and application. Essential oils extracted from plants contain a variety of organic compounds with corrosion inhibition properties, known for their green and sustainable characteristics [43–45]. Challenges include volatility, compatibility, and cost. However, essential oils are considered eco-friendly and often biodegradable, and their natural origins align with the green and sustainable corrosion inhibition approach. Agriculture, biomass, biopolymer wastes, and essential oils hold significant promise as green corrosion inhibitors due to their abundant, renewable, and eco-friendly nature. Ongoing research and development efforts are essential to optimize these green inhibitors and unlock their full potential for sustainable corrosion protection.

6.4.3 Limitations of Research Techniques Utilized in Corrosion Inhibition Investigations

The development of effective environmentally sustainable corrosion inhibitors relies on reliable and precise research techniques. However, existing research methods have limitations that hinder comprehensive understanding and optimization of corrosion inhibitors. These limitations include simplified laboratory conditions, limited insight into mechanisms, surface sensitivity, static analysis, and computational resources [46–49]. Electrochemical methods, such as potentiodynamic polarization and electrochemical impedance spectroscopy (EIS), are often conducted in simplified laboratory environments, which do not fully represent the complex and dynamic conditions encountered in real-world corrosion scenarios. This discrepancy can lead to discrepancies between laboratory results and actual field performance [50–52].

Surface analysis techniques, such as X-ray photoelectron spectroscopy (XPS) and scanning electron microscopy (SEM), are surface-sensitive and provide information about the outermost atomic layers, but they may not capture the full extent of inhibitor–metal interactions occurring beneath the surface. Researchers are exploring in situ and real-time surface analysis techniques to observe corrosion inhibition processes and gain a more comprehensive understanding of inhibitor behavior [53, 54]. Molecular modeling and simulation require substantial computational resources and expertise, limiting their accessibility for many researchers and organizations. Accuracy of models depends on the quality of input parameters and force field parameters, which can lead to discrepancies between predicted and experimental results. Advances in computational chemistry are continually improving the accuracy and efficiency of molecular modeling techniques, and efforts are underway to develop user-friendly software tools for inhibitor design and evaluation.

Environmental considerations are also limited by existing research techniques, as they often do not comprehensively evaluate the environmental impact of corrosion inhibitors [55]. Researchers are increasingly integrating green chemistry principles into inhibitor development, conducting life cycle assessments, and considering the entire lifecycle of inhibitors from production to disposal to assess their environmental impact. Innovative approaches in research techniques encompass machine learning and artificial intelligence, advanced spectroscopy, corrosion sensors, nanotechnology, and green screening methods [31]. These advancements hold the potential to revolutionize corrosion inhibition research by providing more accurate, efficient, and sustainable solutions for protecting critical infrastructure and assets.

6.5 SUMMARY

The global corrosion inhibitor market is experiencing significant growth due to the expansion of industries such as oil and gas, automotive, infrastructure, and manufacturing. The demand for sustainable corrosion inhibition solutions has never been greater, and this market faces challenges such as inconsistent composition and limited effectiveness in waste extracts, agriculture and biomass waste, biopolymer waste, and essential oils.

Marketplace dynamics are influenced by technological advancements, regulatory requirements, and new markets, with regulatory mandates promoting safer and more sustainable solutions. Green corrosion inhibitors derived from waste extracts, agriculture, biomass, biopolymer waste, and essential oils signify a promising frontier for eco-friendly alternatives that reduce environmental and health risks. However, these eco-friendly alternatives face challenges such as inconsistent composition, formulation complexity, film formation and solubility issues, and volatility and compatibility issues.

Effective corrosion inhibitor development relies on robust research techniques, but there are limitations in electrochemical methods, surface analysis techniques, molecular modeling, and environmental considerations. Innovative approaches, such as machine learning, advanced spectroscopy, corrosion sensors, nanotechnology, and green screening, are emerging to overcome these limitations.

In conclusion, the global corrosion inhibitor market is at the crossroads of growth and sustainability, driven by the need for environmentally friendly corrosion protection solutions. Addressing these challenges, advancing research techniques, and fostering innovation will enable the commercialization of sustainable corrosion inhibitors, allowing industries to safeguard their assets, reduce environmental impact, and align with global goals of sustainability and responsible resource management.

REFERENCES

1. Leygraf, C., Wallinder, I.O., Tidblad, J. and Graedel, T., 2016. *Atmospheric Corrosion*. Wiley.
2. Thakur, A. and Kumar, A., 2021. Sustainable inhibitors for corrosion mitigation in aggressive corrosive media: A comprehensive study. *Journal of Bio-and Tribo-Corrosion*, *7*, pp.1–48.
3. Fayomi, O.S.I., Akande, I.G. and Odigie, S., 2019, December. Economic impact of corrosion in oil sectors and prevention: An overview. *Journal of Physics: Conference Series*, *1378*(2), 022037.
4. Alamri, A.H., 2020. Localized corrosion and mitigation approach of steel materials used in oil and gas pipelines – An overview. *Engineering Failure Analysis*, *116*, 104735.
5. Habib, S., Shakoor, R.A. and Kahraman, R., 2021. A focused review on smart carriers tailored for corrosion protection: Developments, applications, and challenges. *Progress in Organic Coatings*, *154*, 106218.
6. Bender, R., Féron, D., Mills, D., Ritter, S., Bäßler, R., Bettge, D., De Graeve, I., Dugstad, A., Grassini, S., Hack, T., Halama, M., Han, E.H., Harder, T., Hinds, G., Kittel, J., Krieg, R., Leygraf, C., Martinelli, L., Mol, A., Neff, D., Nilsson, J.O., Odnevall, I., Paterson, S., Paul, S., Prošek, T., Raupach, M., Revilla, R.I., Ropital, F., Schweigart, H., Szala, E., Terryn, H., Tidblad, J., Virtanen, S., Volovitch, P., Watkinson, D., Wilms, M., Winning, G. and Zheludkevich, M., 2022. Corrosion challenges towards a sustainable society. *Materials and Corrosion*, *73*(11), pp.1730–1751.
7. Betti, N., Al-Amiery, A.A. and Al-Azzawi, W.K., 2022. Experimental and quantum chemical investigations on the anticorrosion efficiency of a nicotinehydrazide derivative for mild steel in HCl. *Molecules*, *27*(19), p.6254.
8. Vazquez, O., 2023. Scale inhibitors. In *Modelling Oilfield Scale Squeeze Treatments: From Core to Reservoir* (pp.35–55). Springer International.
9. Verma, C., Ebenso, E.E., Quraishi, M.A. and Hussain, C.M., 2021. Recent developments in sustainable corrosion inhibitors: Design, performance and industrial scale applications. *Materials Advances*, *2*(12), pp.3806–3850.
10. Defteraios, N., Kyranoudis, C., Nivolianitou, Z. and Aneziris, O., 2020. Hydrogen explosion incident mitigation in steam reforming units through enhanced inspection and forecasting corrosion tools implementation. *Journal of Loss Prevention in the Process Industries*, *63*, 104016.

11. Palanisamy, G., 2019. Corrosion inhibitors. In *Corrosion Inhibitors* (pp.1–24). IntechOpen.
12. Aslam, R., Mobin, M., Zehra, S. and Aslam, J., 2022. A comprehensive review of corrosion inhibitors employed to mitigate stainless steel corrosion in different environments. *Journal of Molecular Liquids*, *364*, 119992.
13. Haldhar, R., Kim, S.C., Berdimurodov, E., Verma, D.K. and Hussain, C.M., 2021. Corrosion inhibitors: Industrial applications and commercialization. In *Sustainable Corrosion Inhibitors II: Synthesis, Design, and Practical Applications* (pp.219–235). American Chemical Society.
14. Reza, N.A., Akhmal, N.H., Fadil, N.A. and Taib, M.F.M., 2021. A review on plants and biomass wastes as organic green corrosion inhibitors for mild steel in acidic environment. *Metals*, *11*(7), p.1062.
15. Singh, A.K., 2023. Importance and contribution of carbon allotropes in a green and sustainable environment. In *Carbon Allotropes and Composites: Materials for Environment Protection and Remediation* (pp.337–382). Wiley.
16. Chauhan, D.S., Quraishi, M.A. and Qurashi, A., 2021. Recent trends in environmentally sustainable Sweet corrosion inhibitors. *Journal of Molecular Liquids*, *326*, 115117.
17. Shang, Z. and Zhu, J., 2021. Overview on plant extracts as green corrosion inhibitors in the oil and gas fields. *Journal of Materials Research and Technology*, *15*, pp.5078–5094.
18. Zunita, M. and Kevin, Y.J., 2022. Ionic liquids as corrosion inhibitor: From research and development to commercialization. *Results in Engineering*, *15*, 100562.
19. Al-Amiery, A.A. and Al-Azzawi, W.K., 2023. Mannich bases as corrosion inhibitors: An extensive review. *Journal of Molecular Structure*, *7*, 136421.
20. Saxomiddino'g'li, Z.S. and Marg'uba, Y., 2023. Protecting metals from corrosion: Preserving durability and longevity. *American Journal of Engineering, Mechanics and Architecture (2993–2637)*, *1*(5), pp.11–13.
21. Medupin, R.O., Ukoba, K., Yoro, K.O. and Jen, T.C., 2023. Sustainable approach for corrosion control in mild steel using plant-based inhibitors: A review. *Materials Today Sustainability*, *22*, 100373.
22. Alao, A.O., Popoola, A.P. and Sanni, O., 2022. The influence of nanoparticle inhibitors on the corrosion protection of some industrial metals: A review. *Journal of Bio-and Tribo-Corrosion*, *8*(3), p.68.
23. Khan, M.A.A., Irfan, O.M., Djavanroodi, F. and Asad, M., 2022. Development of sustainable inhibitors for corrosion control. *Sustainability*, *14*(15), p.9502.
24. Aslam, J., Verma, C. and Aslam, R., Eds., 2023. *Grafted Biopolymers as Corrosion Inhibitors: Safety, Sustainability, and Efficiency*. Wiley.
25. Ji, D., 2020. Growth of graphene as anti-corrosive coating. Doctoral dissertation, UNSW, Sydney.
26. Avtar, R., Tripathi, S., Aggarwal, A. K. and Kumar, P., 2019. Population–urbanization–energy nexus: A review. *Resources*, *8*(3), p.136.
27. Haddaoui, I. and Mateo-Sagasta, J., 2021. A review on occurrence of emerging pollutants in waters of the MENA region. *Environmental Science and Pollution Research*, *28*(48), pp.68090–68110.
28. Anderson, R.D., Jones, A. and Kovacic, W.E., 2018. Preventing corruption, supplier collusion, and the corrosion of civic trust: A procompetitive program to improve the effectiveness and legitimacy of public procurement. *George Mason Law Review*, *26*, p.1233.

29. Liao, J.J., Abd Latif, N.H., Trache, D., Brosse, N. and Hussin, M.H., 2020. Current advancement on the isolation, characterization and application of lignin. *International Journal of Biological Macromolecules*, *162*, pp.985–1024.
30. Salleh, S.Z., Yusoff, A.H., Zakaria, S.K., Taib, M.A.A., Seman, A.A., Masri, M.N., Mohamad, M., Mamat, S., Sobri, S.A., Ali, A. and Teo, P.T., 2021. Plant extracts as green corrosion inhibitor for ferrous metal alloys: A review. *Journal of Cleaner Production*, *304*, 127030.
31. Ebenso, E.E., Verma, C., Olasunkanmi, L.O., Akpan, E.D., Verma, D.K., Lgaz, H., Guo, L., Kaya, H. and Quraishi, M.A., 2021. Molecular modelling of compounds used for corrosion inhibition studies: A review. *Physical Chemistry Chemical Physics*, *23*(36), pp.19987–20027.
32. Perera, M., Yan, J., Xu, L., Yang, M. and Yan, Y., 2022. Bioprocess development for biolubricant production using non-edible oils, agro-industrial byproducts and wastes. *Journal of Cleaner Production*, *357*, 131956.
33. Gołębiewska, E., Kalinowska, M. and Yildiz, G., 2022. Sustainable use of apple pomace (AP) in different industrial sectors. *Materials*, *15*(5), p.1788.
34. More, P., Jangam, K., Gardi, S., Athavale, R., Choudhary, F. and Yamgar, R., 2023. *Sustainable Grafted Biopolymers. Grafted Biopolymers as Corrosion Inhibitors: Safety, Sustainability, and Efficiency* (pp.89–120). Wiley.
35. Tayyab, M., Noman, A., Islam, W., Waheed, S., Arafat, Y., Ali, F., Zaynab, M., Lin, S., Zhang, H. and Lin, W., 2018. Bioethanol production from lignocellulosic biomass by environment-friendly pretreatment methods: A review. *Applied Ecology & Environmental Research*, *16*(1), pp.225–249.
36. Mwakalesi, A.J. and Nyangi, M., 2022. Effective corrosion inhibition of mild steel in an acidic environment using an aqueous extract of macadamia nut green peel biowaste. *Engineering Proceedings*, *31*(1), p.41.
37. Kumar, P., Barrett, D.M., Delwiche, M.J. and Stroeve, P., 2009. Methods for pretreatment of lignocellulosic biomass for efficient hydrolysis and biofuel production. *Industrial & Engineering Chemistry Research*, *48*(8), pp.3713–3729.
38. Zhou, C.H., Xia, X., Lin, C.X., Tong, D.S. and Beltramini, J., 2011. Catalytic conversion of lignocellulosic biomass to fine chemicals and fuels. *Chemical Society Reviews*, *40*(11), pp.5588–5617.
39. Marzorati, S., Verotta, L. and Trasatti, S.P., 2018. Green corrosion inhibitors from natural sources and biomass wastes. *Molecules*, *24*(1), p.48.
40. Alrefaee, S.H., Rhee, K.Y., Verma, C., Quraishi, M.A. and Ebenso, E.E., 2021. Challenges and advantages of using plant extract as inhibitors in modern corrosion inhibition systems: Recent advancements. *Journal of Molecular Liquids*, *321*, 114666.
41. Munawaroh, H.S.H., Sunarya, Y., Anwar, B., Priatna, E., Risa, H., Koyande, A.K. and Show, P.L., 2022. Protoporphyrin extracted from biomass waste as sustainable corrosion inhibitors of T22 carbon steel in acidic environments. *Sustainability*, *14*(6), p.3622.
42. Mobin, M., Cial, K., Aslam, J., Parveen, M. and Aslam, R., 2023. Grafted dextrin as a corrosion inhibitor. In *Grafted Biopolymers as Corrosion Inhibitors: Safety, Sustainability, and Efficiency* (pp.383–396). Wiley.
43. Vorobyova, V.I., Skiba, M.I., Shakun, A.S. and Nahirniak, S.V., 2019. Relationship between the inhibition and antioxidant properties of the plant and biomass wastes extracts – A review. *International Journal of Corrosion and Scale Inhibition*, *8*(2), pp.150–178.
44. Daoudi, W., El Aatiaoui, A., Falil, N., Azzouzi, M., Berisha, A., Olasunkanmi, L.O., Dagdag, O., Ebenso, E.E., Koudad, M., Aouinti, A., Loutou, M. and Oussaid, A., 2022.

Essential oil of Dysphania ambrosioides as a green corrosion inhibitor for mild steel in HCl solution. *Journal of Molecular Liquids, 363*, 119839.

45. Zakeri, A., Bahmani, E. and Aghdam, A.S.R., 2022. Plant extracts as sustainable and green corrosion inhibitors for protection of ferrous metals in corrosive media: A mini review. *Corrosion Communications*, 5, pp.25–38.
46. Guo, L., Obot, I.B., Zheng, X., Shen, X., Qiang, Y., Kaya, S. and Kaya, C., 2017. Theoretical insight into an empirical rule about organic corrosion inhibitors containing nitrogen, oxygen, and sulfur atoms. *Applied Surface Science, 406*, pp.301–306.
47. Singh, A., Ansari, K.R., Chauhan, D.S., Quraishi, M.A., Lgaz, H. and Chung, I.M., 2020. Comprehensive investigation of steel corrosion inhibition at macro/micro level by ecofriendly green corrosion inhibitor in 15% HCl medium. *Journal of Colloid and Interface Science*, 560, pp.225–236.
48. Askari, M., Aliofkhazraei, M., Jafari, R., Hamghalam, P. and Hajizadeh, A., 2021. Downhole corrosion inhibitors for oil and gas production – A review. *Applied Surface Science Advances, 6*, 100128.
49. Jin, H., Wang, J., Tian, L., Gao, M., Zhao, J. and Ren, L., 2022. Recent advances in emerging integrated antifouling and anticorrosion coatings. *Materials & Design, 213*, 110307.
50. Jadhav, N. and Gelling, V.J., 2019. The use of localized electrochemical techniques for corrosion studies. *Journal of the Electrochemical Society, 166*(11), C3461.
51. Trentin, A., Pakseresht, A., Duran, A., Castro, Y. and Galusek, D., 2022. Electrochemical characterization of polymeric coatings for corrosion protection: A review of advances and perspectives. *Polymers, 14*(12), p.2306.
52. Kumari, P. and Kumar, A., 2023. In-depth analysis of corrosion-resistance and dynamic stability of superhydrophobic copper developed by a green approach. *Materials Today Communications, 36*, 106744.
53. Wang, J., Du, M., Li, G. and Shi, P., 2022a. Research progress on microbiological inhibition of corrosion: A review. *Journal of Cleaner Production, 373*, 133658.
54. Kozlica, D.K., Izquierdo, J., Souto, R.M. and Milošev, I., 2023. Inhibition of the localised corrosion of AA2024 in chloride solution by 2-mercaptobenzimidazole and octylphosphonic acid. *npj Materials Degradation, 7*(1), p.55.
55. Prasad, A.R., Kunyankandy, A. and Joseph, A., 2020. Corrosion inhibition in oil and gas industry: Economic considerations. In *Corrosion Inhibitors in the Oil and Gas Industry* (pp.135–150). Wiley.

7 Corrosion and Emerging Technologies

7.1 INTRODUCTION TO EMERGING TECHNOLOGY AND 4IR

Emerging technology is any fast-growing, ambiguous technology marked with a high degree of coherence that is persistent and impacts the socio-economic domain [1]. The fourth industrial revolution (4IR) began in 2016 when the chairman of the World Economic Forum, Klaus Schwab, coined the phrase. 4IR is defined as the advent of "exceptional technological improvements" that combine the physical, biological, social, and digital realms. Since the advent of the fourth industrial revolution, certain technologies have emerged that suit the definition of emerging technology based on the ambiguity, coherence, and impact on the socio-economic of the world.

Emerging technologies can be grouped on the era, region, timeline, and technology as depicted in Figure 7.1.

Emerging technology has been published in over 2,200 publications according to Rotolo, Hicks and Martin [2]. Approximately 50% of these publications focus on emerging technology in the educational sector and 12% focus on science and technology.

7.2 EFFECT OF CORROSION ON EMERGING TECHNOLOGY

Corrosion is a billion-dollar industry that affects all sectors. Emerging technologies are not spared. Most commercial planes and electric cars are also at the receiving end of damage caused by corrosion. Cloud computing also suffers from corrosion caused by damage to data centers, and server rooms, among others. Corrosion will continue to impact technologies and by extension humanity.

7.3 KEY EMERGING TECHNOLOGIES

Spintronics, robotics, biotechnology, artificial intelligence (AI), nanotechnology, gene therapy, three-dimensional (3D) printing, stem cell therapy, and multi-use rockets are some of the emerging technologies of the last decade.

Technologies such as wireless charging of low-powered devices, breath sensors capable of diagnosing diseases, on-demand drug manufacturing, self-fertilizing crops, and 3D printing are some of the top 10 key emerging technologies predicted

DOI: 10.1201/9781003441151-7

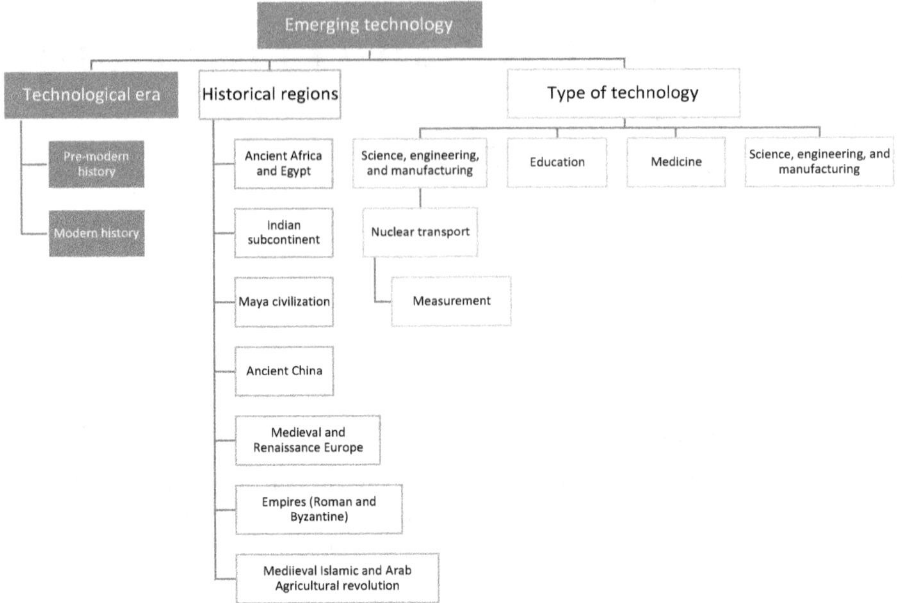

FIGURE 7.1 Classification of emerging technology.

from 2020 to 2021 to impact the world in the next 5 years as compiled by the World Economic Forum.

7.3.1 Self-Fertilizing Crop

The growing population requires more production of food to ensure human existence. This has led to demand for fertilizers to produce food. About 100 million tons of nitrogen fertilizers are used to improve crop yield yearly. Hence, there is an innovation that makes fertilizers for legumes, corn, and cereals by making crops to produce the nitrogen captured from ammonia. Another form of self-fertilizing is the production of nitrogenase by soil bacteria to convert atmospheric nitrogen into ammonia for plant use. This chapter aims to spell out technology and methods to ensure providing an extensively available and steady supply of food for humanity.

7.3.2 Breath Sensors Diagnose Diseases

This technology will help people who are averse to needles for drawing blood. The technology uses a breath analyzer to detect diseases. The breath sensor determined the disease by sampling the concentration of over 800 compounds contained in human breath against a pre-installed database. The sensor uses the change in electrical resistance caused by the flow of breath over a metal-oxide-semiconductor to generate data. Thereafter, an algorithm then ascertained the type of disease from the data generated.

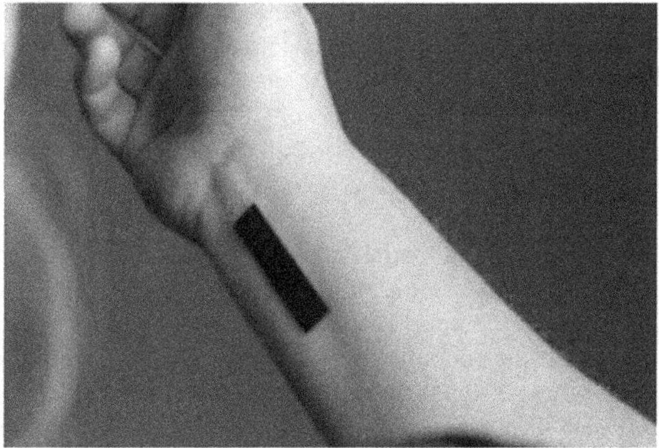

FIGURE 7.2 Wireless biomarker for data on diabetes.

A high concentration of acetone in human breath shows the possible presence of diabetes mellitus. The model name of this breath sensor is *breathonix*.

A similar technology deployed by over 100 companies worldwide uses a wireless biomarker as shown in Figure 7.2 to provide data on diabetes from patients. This is useful for diabetes and requires constant needles. This wireless biomarker obviates the need for needles.

7.3.3 On-Demand Drug Manufacturing

This technology ensures that custom-made drugs can be produced irrespective of the location for those who can afford the high cost. Conventional medicines are made in batch systems, multistep with different parts dispersed in regions across the globe. This usually takes a longer period, and more materials and produced for general usage. However, this technology eliminates all that and can even produce drugs for a single patient. It moves the drug materials through tubes into small reaction chambers.

7.3.4 Self-Healing Materials

Imagine a bad road filled with potholes being able to repair itself continuously, and a torn fabric is able to repair itself. That is what self-healing materials offer. These materials partially or fully restore to their original properties when damaged. Examples of self-healing materials include metals, polymers, ceramics, and their composites [3]. Others are nanoparticles, graphene, nanotubes, self-healing polymers, polyurethane, polymerization, hydrogel, and intelligent material. The principle employed by self-healing materials is akin to wound healing in which damage is restored to its original state within minutes, hours, or days depending on the material and the damage. The technology is bioinspired and is capable of healing micro or nano-level cracks in such materials with any intervention. Self-healing can be achieved via covalent bonding and molecular heterogeneities for the repair process. Self-healing involves the reformation of certain bonds in the materials. The bonds are reversible covalent and non-covalent bonds. In a reversible covalent bond, a molecular heterogeneity is introduced into a polymer backbone that results in the formation of new phases that initiate the repair. Nanocomposites' interfacial regions also provide a platform for supramolecular assemblages or covalent bonding, which, with the right design, can self-heal.

Self-healing has been used in clothes and other applications. The squid ring teeth (SRT) polymer used for self-healing cloth was developed by the Demirel laboratory in Penn State [4]. Squid ring teeth are structural proteins having self-healing properties, and high elastic modulus in wet and dry states due to semicrystalline arrangement. The clothes self-heal when in contact with water making the cloth wearable and reusable and absorbing toxins. The cloth contains organophosphate hydrolase that breaks down the toxins. The coating is thin, and unnoticeable and improves the strength of the cloth. Research is going on to use natural and synthetic materials for self-healing. Lee and Park [5] used mushroom-derived chitinous polymers to produce self-healing polymers.

7.3.5 Energy from Wireless Signal

Would not it be nice if you could charge your device from Wi-Fi signal? That is what this technology does. This technology generates energy from radio-frequency (RF) signals by converting AC electromagnetic to direct current (DC) electricity to power devices. The energy generated is used to charge batteries or power devices such as those for low-power sensor networks, RFID, the Internet of Things, and other devices.

7.3.5.1 How Does it Work

An oscillating magnetic and electric field is created by the wire in an RF antenna caused by alternating of the negative and positive charge. These fluctuating energies radiate as electromagnetic waves move toward space and can be picked up by a device nearby. Different studies have been done to advance this technology over the years [6–8].

7.3.6 Body-Adapted Wearable Electronic

This technology dates back to 2014. This technology monitors lactate, glucose, and heart rate using devices worn by humans. Wearable devices such as pants, smart watches, health trackers, and other devices. The flexibility offered by flexible electronics allows integration into wearable devices. This allows them to operate in the microwave, millimeter wave range. Wearables can be used to detect heart rate, movement, calories burned, blood pressure, seizures, exercise routine, physical strain, and biochemicals, among others. It is used in fashion (clothing and accessories), entertainment (virtual and augmented reality), military (training, and monitoring health and physiological of soldiers), and gaming (pong, video game console).

There are currently about five types that exist. They are smart watches, smart hearables, smart clothing, smart implantables, and smart patches.

Smart watches are trendy watches worn on the wrist with additional features to the conventional watches. They come in various sizes, shapes, and feel with the functionality varying depending on the brand and the purchasing power. The most common features include fitness tracking, heart rate monitoring, health tracking, SMS, and calls received in tandem with a mobile phone. Some watches also control other functionalities of the connected mobile phone such as music, app among others. The market share of smart watches was around 17 billion USD in 2019. There is a projected growth rate of about 11.6% for 2024.

Smart hearables are among the 2022 trends of wearables with an expected global market of around 19.8 billion USD by 2025. The about 35.4% predicted increment within the period is driven by huge demand for technological advancement in headsets and those in need of hearing aids. Smart hearable devices infuse AI and advanced sensors into existing headset technology to give users a customized and optimized experience. Google, Amazon, Bose, and Apple are leading in the deployment of this technology.

Smart clothing is a technology that infuses sensors and electronics into fabrics. Smart clothing is defined as articles of clothing with sensors that can gather, store, and exchange data from the body or the surroundings. Users can view the data after it has been transferred via Bluetooth to apps on their smartphones. Conversely, "smart clothes" only refer to fashionable items like headwear, shorts, swimsuits, and t-shirts (like Bluetooth beanies). However, both consumers and businesses frequently confuse the word with smart clothing. Small generators or batteries are used to convert body heat to electricity. The smart clothing tracks skin temperature, heart rate, respiration, and other biometric data using sensors. It is also used to control cooling and heating in houses without the need for a phone. It is used for virtual reality and augmented reality.

Materials varying from conventional cotton, polyester, and nylon to cutting-edge Kevlar with built-in functionalities can be used to create smart textile fabric. However, at this time, electrically conductive fabrics are of concern. Smart clothing examples include spacesuits, musical jackets, I-wear, data wear, sports jackets, intelligent bras, and wearable computers.

Smart clothing had a market share of 1.4 billion USD in 2020. This is expected to grow to 2.8 billion by 2025 driven by advancements in textiles, and growing interest in wearable devices and fitness.

Smart implantable is a technology infused into the body. This technology interfaces with organs and tissues to diagnose, monitor, and treat ailments. They produce titanium, polymers, and stainless steel that are biocompatible to the human body. They could be active or passive. The difference is that passive device uses energy from the body without any electronics. The active smart implantable uses a battery and electronic device. The global market is driven by chronic ailments, rising demand for invasive surgeries, and an aging population with an expected growth projected to be 32 USD by 2025.

Smart patches are thin and flexible devices attached to the skin to monitor health parameters and deliver medication. They are infused with sensors for obtaining physiological parameters. Parameters such as body temperature, heart rate, and blood oxygen levels are monitored by smart patches and alerts for divergence from original functioning. In some cases, the smart patches have microchips that deliver drugs. Diabetic patients can use it to deliver insulin. It is also used for remote tracking of elderly person health. The smart patches market is expected to grow to about 21% in 2027. This is driven by the growing need for monitoring patients, chronic disease, and non-invasive treatment methods.

7.3.7 Electric Aviation

Electric aviation involves using electricity to power aircraft and other aviation flying equipment. This is in a bit to reduce about 2% greenhouse gas emission by aircraft and ensure quieter flight. It also reduces the need for scheduled maintenance of aircraft as current gasoline aircraft are serviced every 2,000 hours as opposed to 20,000 hours for electric motors. The electricity used is supplied by different methods but the battery is the most common. The first electric aircraft was the PKZ-1 which used tethering wires that connected it to the ground. This flight achieved a height of 50 m.

Electric A modified Cessna Caravan 208B capable of carrying nine passengers and a pilot is the largest electric airplane to take to the air. The aircraft flew for 30 minutes over Grant County International Airport in central Washington state. It is fitted with an electric motor manufactured by MagniX and powered by lithium batteries. However, long-haul flights are still far from being a reality because of energy density. The energy density of jet fuel is around 12,000 Wh/kg and lithium-ion battery energy density peaks at 250 Wh/kg.

Airbus, an aircraft manufacturer, has started leveraging the technology by deploying sustainable aviation fuel (SAF) with electricity. The hybrid technology allows some of the airbuses fitted with the technology to use jet fuel and electricity either simultaneously or separately.

7.3.8 Artificial Intelligence (AI)

It is an emerging field that uses computer science and datasets to solve problems. The term artificial intelligence was coined in the 1950s by John McCarthy. This includes machine and deep learning. About four types of artificial intelligence exist viz theory of mind, self-awareness, limited memory, and reactive machine. Java and Python are the most popular programming languages used in AI. AI offers a human-like approach to solving problems though unable to replace humans for now. Artificial narrow intelligence (ANI), artificial general intelligence (AGI), and artificial super intelligence (ASI) are the three types of AI. AI is used in various spheres of human endeavor from simple tasks to complex tasks. It is used in mobile phones, stocks, and search optimization among others. The six key technologies of AI include automated speech recognition, computer vision for drones and other applications, natural language processing, AI planning, expert systems, and machine learning.

7.3.9 3D Printing Technology

The technology of 3D printing has been around since but recently started gaining ground as shown in Figure 7.3.

The acceptance and wide application are connected to the great advantage of the technology. It reduces material wastage; it gives cost benefit to parts up to 100. It can be customized with little or limited technical ability. Recently, it has become affordable as there are 3D printers that cost less than 250 USD for filament and 300 USD for resin 3D printers. 3D printers can now be printed using different materials from metals, polymers, rubber, resins, ceramics, and cement, among others.

For countries like the USA and other developed countries, this may not be hugely celebrated. However, in countries in the global south like South Africa, Nigeria, and Kenya where it takes sometimes more than a year to build a simple 3-bedroom apartment by an individual, this is a big plus. The cost of building houses in developing countries is affected by the need to import them, weather, and other factors. 3D printing has resulted in timely and affordable housing for everyone. It eliminates about 95% of the material requiring transportation as it is capable of using local materials and the 3D printer to get everything done at the house site. The materials used are mostly locally sourced and they include sand, clay, and local fibers. It is capable of providing rugged shelter to slums such as Mukuru in Nairobi, Mushin and Ajegunle in Lagos, and Soweto Johannesburg in Kenya, Nigeria, and South Africa, respectively. This technology has been deployed in South Africa with a six-bedroom built in the University of Johannesburg, South Africa, in Malawi, Kenya, and other places. This maybe the game changer in the housing sector for those in the global south. 3D printing has also been deployed for space usage. It is believed to help provide structure in outer space.

In 2017, the business established S-Squared 4D Commercial, a new branch that uses its Autonomous Robotic Construction System 3D printing gear to construct houses and commercial buildings (ARCS) [9]. Robert Smith and Mario Szczepanski co-founded this bootstrapped business, which now employs 13 people. The Autonomous Robotic Construction System (ARCS) is a 20-by-40-foot environmentally friendly concrete printer that can construct a 1,490-square-foot home in 36 hours. The system can construct houses, businesses, roads, and bridges. ARCS can handle projects ranging in size from 500 square feet to over one million square feet. Mario Cucinella

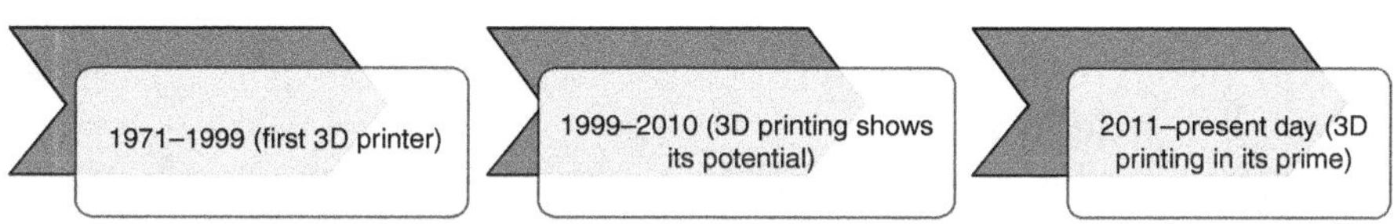

FIGURE 7.3 History of 3D printing.

Architects and WASP, a 3D printing company, showed the first 3D printing of a house constructed of a clay mixture called Tecla [10, 11].

7.4 CLEAN ENERGY

Clean energy is energy sources that are renewable and do not pollute the environment. The global energy potential of clean energy is depicted in Figure 7.4.

This is a plot adapted from data taken from Guerrero-Lemus and Martínez-Duart [12]. The clean energy potential of conventional energy is less than that of the smallest clean energy (hydropower).

Classification of clean energy: Clean energy can be classified based on the source, and era as shown in Figure 7.5.

7.4.1 Mainstream Clean Energy

These are developed and globally deployed energy sources that are renewable, do not pollute the environment and are sustainable. Wind and solar energy are fastest growing with a contribution of 49 and 45 GW in 2014. However, hydropower is the most competitive clean energy with a key demerit being the location of the plant.

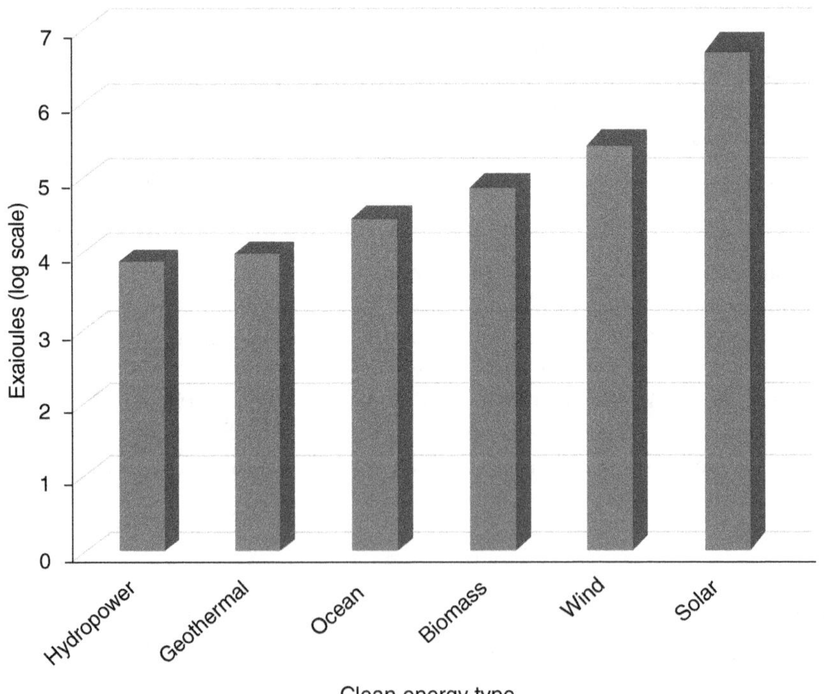

FIGURE 7.4 Worldwide clean energy potential.

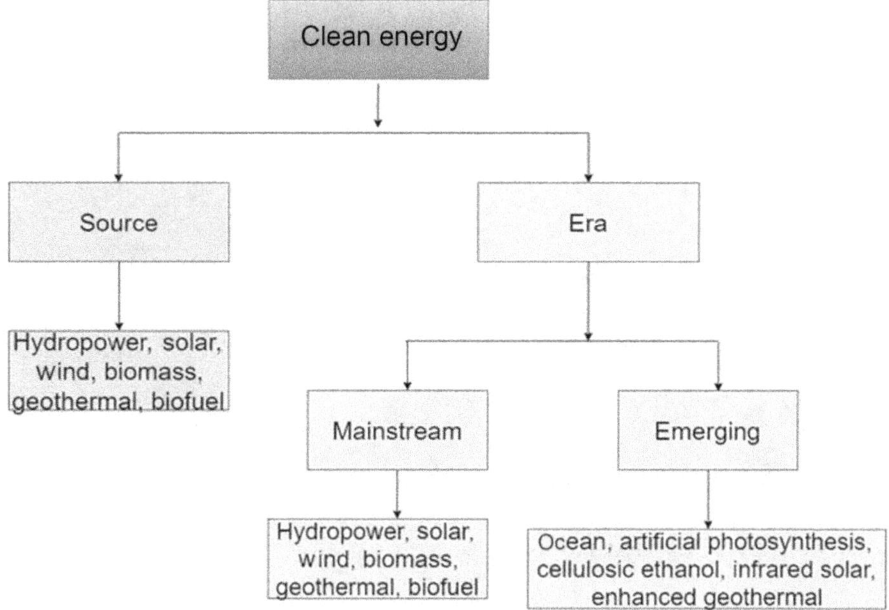

FIGURE 7.5 Classification of clean energy.

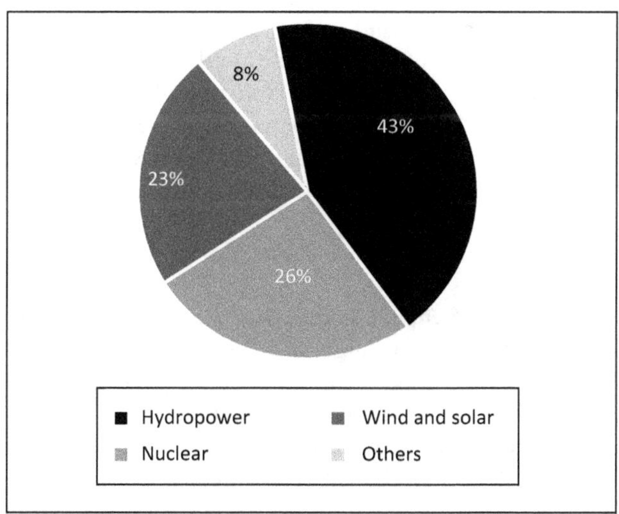

FIGURE 7.6 Low-carbon electricity generation by technology.

Figure 7.6 shows the low-carbon electricity generation for hydropower, nuclear, wind, and solar for 2020. It is a plot of data obtained from the International Energy Agency.

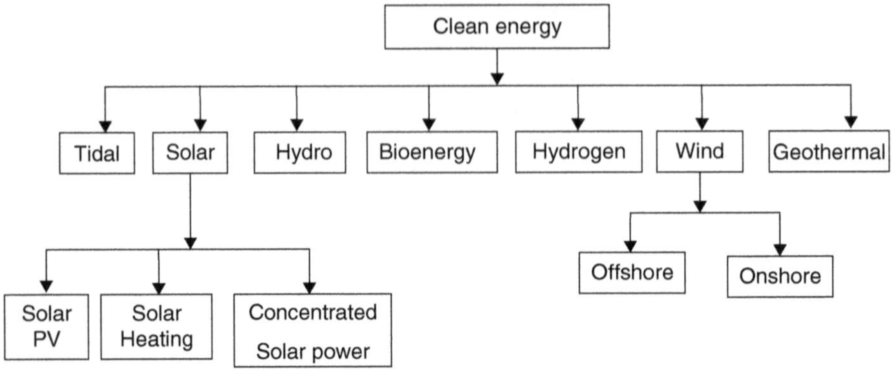

FIGURE 7.7 Classification of clean energy sources.

They include wind energy, hydropower, solar energy, geothermal, and biofuel as shown in Figure 7.7.

They are briefly discussed as follows.

7.4.1.1 Wind Energy

This is a sustainable, clean, popular source of energy that uses wind to turn a generator. Wind converts kinetic energy produced by the wind to electricity. Three sources of wind have been harnessed for electricity generation, which include offshore wind, land-based wind, and distributed wind energy.

A wind farm is a collection of wind turbines in the same place. It could be up to several hundreds of wind turbines spread over an extended area. The adjoining land can be used for farming or to serve other purposes. Horizontal axis wind turbine is the most common design used for large wind farms. They have an upwind rotor of three blades attached to a nacelle to a tall tubular tower. Individual wind turbines are connected with a medium voltage of 34.5 kV with a distance of 7 multiplied by the rotor diameter of the wind turbine. The global onshore wind farms are shown in Figure 7.8.

The highest wind energy farm is found in China; it is known as Gansu with a capacity of 7,965 MW while India has large onshore wind farms known as Muppandal wind farm and Jaisalmer wind farm with both contributing about 2,560 MW. The Gansu wind farms have several thousands of wind turbines.

Total offshore global wind energy capacity stood at 32,000 MW in 2021 with Hornsea Wind Farm in the United Kingdom having the largest offshore wind energy having 1,218 MW capacity.

The global wind energy capacity from 2010 to 2021 is shown in Figure 7.9 as plotted from data obtained from IEA [13].

7.4.1.2 SolarEnergy

Solar energy is the energy derived from the sun. A ray of sunlight falls on a solar panel and excites the electron to move from one region to another resulting in a flow

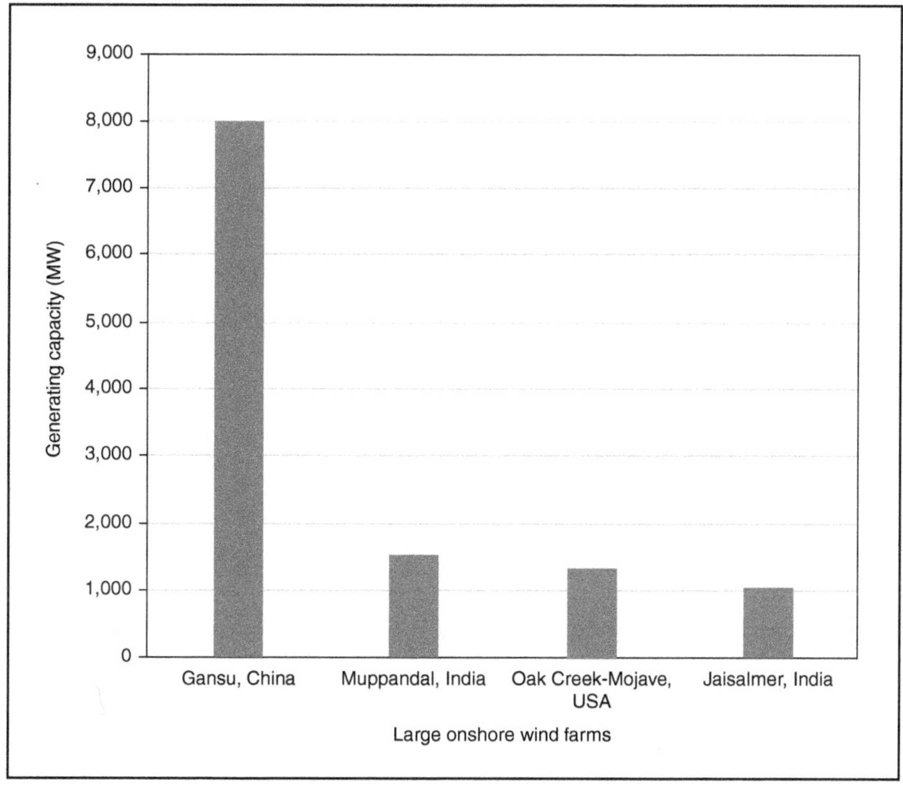

FIGURE 7.8 Large onshore wind farms.

of electricity. The energy from the sun generates heat and electricity. Solar energy is a result of nuclear fusion. The collision of the sun's core with hydrogen protons results in the creation of helium atoms. It is a proton-to-proton chain reaction that occurs in stars as large as the sun resulting in the production of huge energy and heat. About 620 million metric tonnes of hydrogen fuses every second with the sun. The star's temperature is about 4 million Celsius.

Solar cells are the building blocks of solar panels. A large array of solar panels is known as a solar farm. Solar heating, photovoltaic, and concentrating solar power are the three ways used to tap energy from the sun as shown in Figure 7.10.

Solar heating uses heat generated from the sun to provide heating to cold regions of the world. Concentrated solar power is used to power turbines and other devices using energy from the sun. A photovoltaic cell converts the energy from the sun into electricity by using semiconductor materials.

Large-scale applications of the photoelectric effect can be found in solar photovoltaic (solar PV) electricity plants. The material's atoms do not allow the electromagnetic wave-energized electrons to flee. At a particular voltage, the valence electrons can easily move and generate a DC. This electric power is converted to medium voltage (MV –6.6 to 33kV) AC by a large number of solar cells organized in panels

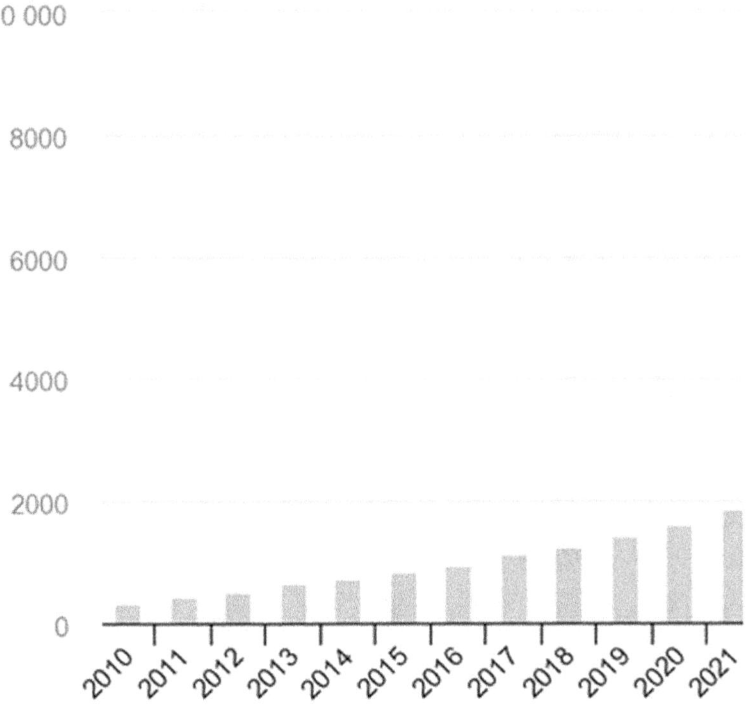

FIGURE 7.9 Global wind energy potential from 2010 to 2021 (IEA, 2022).

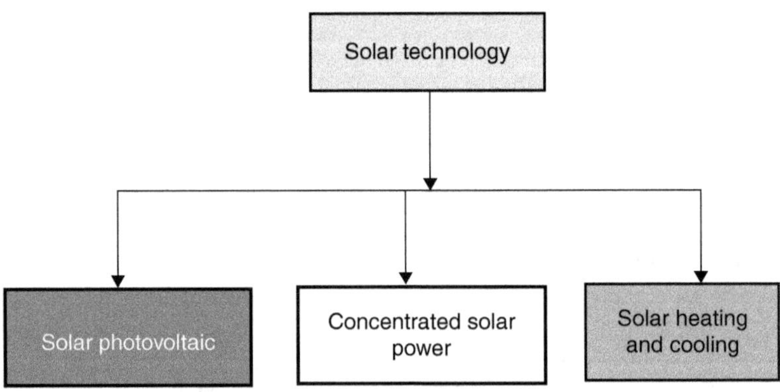

FIGURE 7.10 Type of solar energy technology.

and then strings (AC). The goal of switching from DC to AC is to reduce heat losses caused by strong wire currents ($Ploss = I^2R$).

The building substation receives the MV AC after that. Prior to being fed into the grid, it is converted once more to acquire the voltage that is needed for electrical grid connection (typically 132, 220, 330, or 400kV) [14]. Based on their rated peak power production, or what is known as "Full Sun" under STP conditions (1kW/m^2 of perpendicular irradiance), solar PV panels are marketed. As a result, the best location for solar PV panels would be perpendicular to the sun's path. However, the setting and performance of the solar PV panels are frequently optimized using tools like PVSyst due to the Earth's rotational behavior on its axis and tilt to the sun throughout the year. Other problems can arise in solar PV power plants as well. Temperature losses can surpass 7%, while soil losses can be as high as 3%. With deployed solar PV power plants only capturing between 18% and 25% of the available sunlight to transform into energy, the meteorological irradiance losses are also quite significant.

Electricity can be produced using solar energy on both large and minor scales. However, the two major drawbacks that restrict its use globally are cost and storage [15]. As a result, scientists are trying to figure out how to make solar cells and panels less expensive so they can fight on price with conventional electricity generation techniques. By creating more effective solar cells, this scale can be achieved. Both careful material selection with appropriate energy gaps that fit the solar spectrum and fine-tuning of the material optoelectronic properties can increase solar cell efficiency. Additionally, a new device was developed using single, ternary, and quaternary semiconductor materials for charge collection and improved solar spectrum usage.

More energy is produced by the sun in an hour (4.3 × 1020 J) than the planet Earth uses in a year (4.1 × 1020 J) [16]. The term "insolation" refers to the solar radiation, of which the planet receives 174 petawatts (PW)/year. Insolation is affected by time, place, geography, the sun's position in relation to other stars, the season, and atmospheric refraction, diffusion, and absorption [17, 18]. The oceans, clouds, and land areas absorb approximately70% of this insolation, absorbing 3,850 ZJ/year of solar energy, while 30% is reflected in space [19]. The energy that the sun incidentally sends to Earth each second is equal to 4 trillion 100-watt light bulbs [20]. Notwithstanding these efforts, the potential of solar energy has not been completely realized, with 100 GW of solar energy used globally in 2012. Germany has the highest installed capacity in the globe and Europe has the most installed capacity overall, accounting for about 31% of global solar energy consumption [21].

The photogeneration of charge carriers in a light-absorbing material and their separation to a conductive contact, which will transfer the electricity, are the two main tasks that a solar cell carries out. For this, polymer- and dye-sensitive solar cells require an electrolyte that can transport the charge from the photo-acceptor to the electrodes. In nanoparticle-based solar cells, the particles must be near to one another to transfer charge directly. Quantum dots (QDs) and nanocrystalline materials have been incorporated, which has greatly improved the total efficiencies of solar cells.

Thin-film solar cells, crystalline silicon-based solar cells, and a more recent innovation that combines the two kinds are the three categories into which solar cells fall. According to the time and materials used in their production, solar cells are typically divided into four stages as depicted in Figure 7.11.

7.4.1.3 Hydropower

It employs the kinetic energy of flowing water to spin turbines, which then revolve an electricity-generating generator. Reservoirs and pumped-storage hydropower

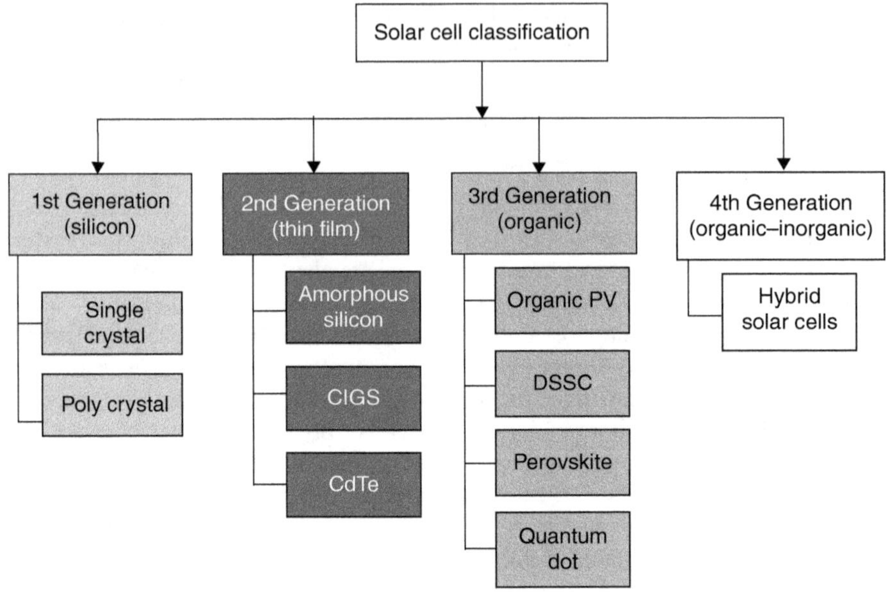

FIGURE 7.11 Solar cell classification based on generation.

stations are the two main techniques used in hydropower stations. Pumped-storage hydroelectric plants cannot run continuously and will not be considered in this case. The head (inlet) and tail (outlet) races are components of a moving river in run-of-river applications. Different mechanisms are employed to run these types of plants, depending on the scale of the installation, the river's flow rate, or the specifications set forth by environmentalists. However, the most basic working principle is to have water collected by the head raceway flow into a Francis-type turbine, where the momentum of the water strikes the blades and causes rotation before being released into the river further downstream from the power station via the tail raceway. Based on the water flow, water head, and the demand for load supply in various seasons, hydropower stations can be divided into different categories as illustrated in Figure 7.12.

China, Canada, and Brazil are the top three hydropower users on the planet. These nations consumed a total of 12.25, 3.59, and 3.42 exajoules (EJ) of hydropower, respectively. With nearly 50%of the world's energy currently coming from hydropower, it forms the foundation of low-carbon electricity production. The output of hydropower is greater than nuclear power by 55% and greater than that of all other renewable energy sources combined, including wind, solar PV, bioenergy, and geothermal. The year 2020 witnessed hydropower contribute 17% of the world's electricity production, making it the third-largest source of energy after coal and natural gas. Globally, hydropower capacity increased by 70% over the past 20 years, but its proportion of total production remained stable as a result of the expansion of wind, solar PV, coal, and natural gas.

Since the 1970s, countries that are emerging or developing have driven the development of the hydropower sector globally, largely as a result of public sector investments

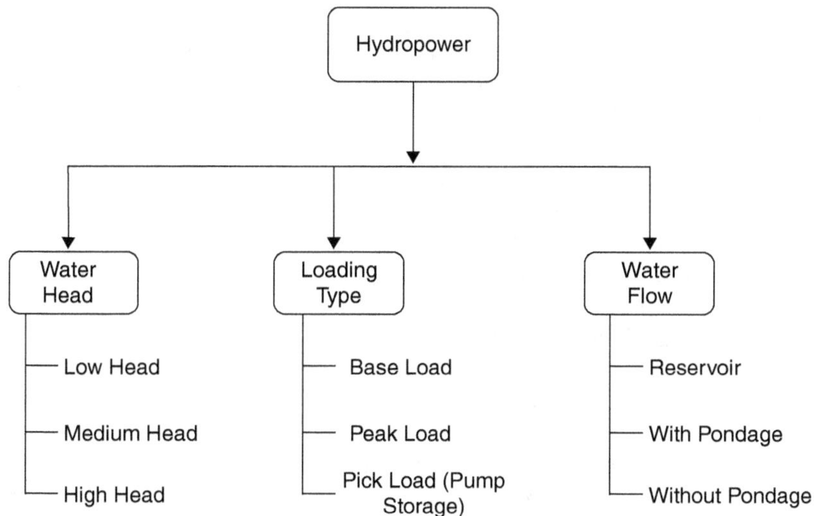

FIGURE 7.12 Classification of hydropower energy.

in large plants. Currently, hydropower supplies the majority of the 800 million people living in 28 emerging and developing countries' electricity needs. It has given those nations a practical means of increasing access to energy. The proportion of hydropower used to generate energy keeps declining in advanced economies because of aging plants. While the typical hydropower plant in Europe is 45 years old, it is almost 50 years old in North America. These outdated fleets, which have been providing on-demand, affordable, and dependable renewable electricity for decades, need to be modernized to continue making a sustainable contribution to electricity security.

Another crucial factor in the flexibility and security of the electricity networks are hydropower plants. Compared to other power plants like nuclear, coal, and natural gas, many hydropower plants can ramp up and scale down their energy production very quickly. Hydropower plants can also be stopped and restarted fairly smoothly. Owing to their high degree of adaptability, they can rapidly respond to changes in demand and make up for variations in the supply of electricity from other sources. Because solar PV and wind energy system's ability to produce electricity can differ depending on various conditions, including the weather, the time of day, and the season, hydropower becomes an attractive option to support their quick deployment and secure integration into electricity systems.

According to our forecast, China will continue to be the world's biggest hydropower market through 2030, contributing 40% to the increase in global capacity. However, since reaching a peak of almost 60% between 2001 and 2010, China's share of worldwide hydropower additions has been decreasing. Because of growing environmental concerns and a shortage of economically appealing locations for big projects, China's hydropower development has slowed down. New long-term goals and financial incentives are expected to unlock a sizable pipeline of previously stalled projects in India, the second-largest development market in the world.

In sub-Saharan Africa, Southeast Asia, and Latin America, more than half of all new hydropower projects are anticipated to be constructed, financed, partly financed, or controlled by Chinese companies by 2030. In sub-Saharan Africa, where it is anticipated that it will be engaged in nearly 70% of new capacity between now and 2030, China's role in hydropower development is the largest. This comprises the Grand Ethiopian Renaissance Dam, the largest hydropower project currently under construction on the planet. A Chinese firm will be involved in the construction of nearly 45% of all hydropower plant capacity in Asia, excluding India, through the year 2030. China is anticipated to make the biggest financial or construction contributions to Pakistan and Lao PDR.

To update outdated plants, mainly in wealthy nations, USD 127 billion, or nearly one-fourth of global hydropower investment, will be spent between now and 2030. Nearly 45% of the hydropower capacity added worldwide over the period will come from work on current facilities, such as the replacement, upgrading, or addition of turbines. The majority of hydropower investment in North America and Europe this decade is expected to go toward modernizing existing facilities. Overall, this investment in modernizing facilities aids in the stability of worldwide hydropower investment relative to the previous 10 years.

7.4.1.4 Geothermal Energy

Geothermal energy is produced by using heat from the Earth's crust or from areas close to heat sources like hot springs or lava ponds. In this case, the typical coal-fired technology arrangement is used yet again, but instead of a boiler, a heat exchanger is used to capture the facility's free natural heat. There is only one geothermal power station in Africa, even though these heat sources are usually reliable. As a result, even though it can provide constant energy, the area is problematic and frequently too expensive to expand.

Geothermal energy is converted to electricity using three main power plant technologies: dry steam, flash steam, and the binary cycle as depicted in Figure 7.13.

The majority of geothermal power plants that are currently in use to produce energy are dry steam or flash plants that use geothermal resources at temperatures higher than 150°C. The development of lower-temperature resources to produce energy or combined heat and power using binary cycle technology, however, is accelerating. When dry steam is generated straight from the geothermal reservoir, dry steam technology is applicable. With this technology, high-pressure saturated or superheated geothermal steam is drawn straight from the geothermal well and used to power a steam turbine and generator . To maximize the effectiveness of energy production, the steam exhaust from the turbine is discharged to a condenser at low pressure or partial vacuum. Backpressure plants that discharge the exhaust steam straight into the atmosphere offer a technologically easier and less expensive option for early electricity production in developing areas in small modular units. Flash steam is the most popular used in geothermal plants. A two-phase geothermal fluid is employed in high temperatures and pressure for electricity production. The two-phase geothermal fluid is vaporized at a lower pressure via a procedure termed "flashing." The geothermal fluid is separated from the liquid component. Thereafter, steam is expanded through

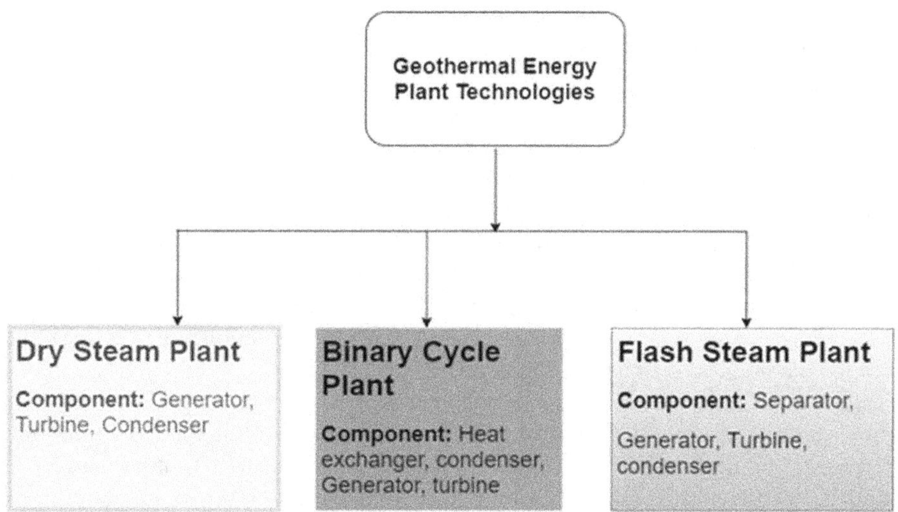

FIGURE 7.13 Classification of geothermal energy plant technologies.

the turbine connected to a generator resulting in electricity generation (single flash). The steam exhaust of the turbine is emptied to a condenser at low pressure or via a backpressure plant solution to the atmosphere. In a double flash, the separated liquid component is flashed twice or thrice, respectively.

Geothermal heating and cooling are grouped into the following applications, namely agriculture and food processing; space heating and cooling; health, recreation and tourism and industrial process heat.

Sedimentary basins and regions with volcanic action both have abundant geothermal resources. Geothermal energy can help stabilize electricity networks with the recent acceleration of the deployment of variable power from wind and solar photovoltaics. A wider range of energy-related uses, such as sustainable heating and cooling, have been made possible by geothermal energy technology, which has advanced beyond its initial emphasis on the electricity market. Geothermal energy is a cost-effective and weather-independent form of renewable energy because of these characteristics. According to IRENA and IGA [22], geothermal energy production has increased at a moderate rate of about 3.5% per year, hitting a total installed capacity of approximately15.96 gigawatts of electricity (GWe) in 2021.

7.4.1.5 Biomass

Energy produced using organic wastes from other activities is known as biomass generation. An illustration of this would be waste produced during the harvest of wheat, as well as treetops, offcuts, and branches left over from forestry operations or the results of deforestation. Because organic biomass can be grown almost forever, it is regarded as a renewable energy source. The energy is generated using the same facilities as conventional coal-fired power plants, with the exception that organic biomass is used as fuel rather than coal. Biomass is burned at the facility to create heat, which is then converted to steam to power a steam turbine and a generator. The procedure is one of the main disadvantages of this energy. However, the carbon in vegetation is oxidized, producing CO_2 emissions.

The future of biomass as a source of sustainable energy is bright. Modern, sustainable biomass technologies are expected to play a significant role in efforts to double the share of renewables in the energy mix, according to REmap 2030, the worldwide roadmap created by the International Renewable Energy Agency (IRENA) [23]. By 2030, biomass use could double from its present level and account for 108 EJ, or 20% of all primary energy supply and 60% of all final renewable energy use, if all the technology choices envisioned in the REmap analysis are implemented. Currently, 35 EJ, or two-thirds of all biomass use, is accounted for by conventional area heating and cooking techniques like burning firewood.

Heat for industry and buildings will get up to 41 EJ, of which only 6 EJ would come from less environmentally friendly conventional uses, accounting for 36 EJ (one-third of the total biomass use in 2030), 36 EJ for power and district heating, and 31 EJ (almost 29%) for transportation. Although there is enough biomass potential in the world to meet the increasing demand, the distribution of various biomass resources is uneven. The possible annual supply of biomass on a global scale is predicted to be between 97 and 147 EJ by 2030. Agricultural waste and residues would account for about 40% of this amount (37–66 EJ). The leftover supply potential is divided between forest products, such as forest residues, and energy crops (33–39 EJ) (24–43 EJ). Geographically, Asia and Europe have the greatest supply capacity, which is calculated to be between 43 and 77 EJ annually. Together, North and South America contribute an additional 45–55 EJ annually.

Africa has the highest yearly potential for energy crops (5–7 EJ); Asia has the highest yearly potential for residues and wastes (15–32 EJ); North America has the lowest yearly potential for energy crops (7 EJ); South America has the highest yearly potential for energy crops (16 EJ); and Europe has the highest yearly potential for fuel wood (0.3–13 EJ) and energy crops (7 EJ).

7.4.2 Emerging Technologies of Clean Energy

There are some emerging technologies in the clean energy sector. They are improvements on existing technologies or exist alone, which are concentrated solar photovoltaics (CSP), ocean energy, cellulosic ethanol, enhanced geothermal energy (EGE), infrared solar cells, and artificial photosynthesis.

7.4.3 Ocean Energy

This is the energy from the ocean and water bodies on earth. About three-quarters of the earth is covered by water sufficient to power the earth [24]. It contains energy in the form of tides, heat, currents, and waves. This vast potential has resulted in research for workable techniques for extracting energy from the ocean. Research has been ongoing on offshore wind. However, some emerging technologies for ocean energy include tidal energy, wave energy, and tide. The key merits of ocean energy are predictability and consistency [25]. Unfortunately, ocean energy currently contributes a minimum amount to global energy [26].

Tidal energy has the highest installed capacity among ocean energy, followed by tides, and salinity gradient as reported by Ocean Energy Systems (OES) in 2015 [27].

7.4.3.1 Ocean Wave Energy

The primary source of this emerging clean energy is the wind that blows in the ocean. It converts the ocean wind into electricity. The ocean wind is consistent, travels long distances with little losses, and is efficient. Temperate latitude of 40° and 60° north and south of the eastern ocean gives the strongest wind for ocean wind energy. The United Kingdom is considered the best for ocean wind energy with other regions theoretical wind energy potential shown in Figure 7.14.

Wave energy extract technology (WEET) is used for extracting ocean wind energy. Yoshi Masuda pioneered the WEET over 200 years ago by developing a navigation buoy. Masuda technology was commercialized in 1965 by Japan, USA [28]. WEET is grouped based on the power take-off and location as depicted in Figure 7.15.

7.4.3.2 Ocean Tidal Energy

This is the energy caused by the movement of ocean waves caused by gravitational and rotational forces between the moon, sun, and earth. The rise and fall of ocean waves are predictable than the sun and wind and regular. However, the technology is still in its infancy, with a small amount of power generated globally. The first commercialized tidal power plant was located in La Rance, France in 1966. The world's largest wind power plant is in South Korea's Sihwa Lake Tidal power station. Nevertheless, there is huge potential for tidal energy in Russia, China, Canada, England, and France. The worldwide potential of tidal energy is above 800 terawatt hours per year and over 40 billion euros per year in global market share.

Types of ocean tidal energy: Tidal barrages, tidal lagoons, and tidal streams are three major ways of generating power from tidal energy as depicted in Figure 7.16.

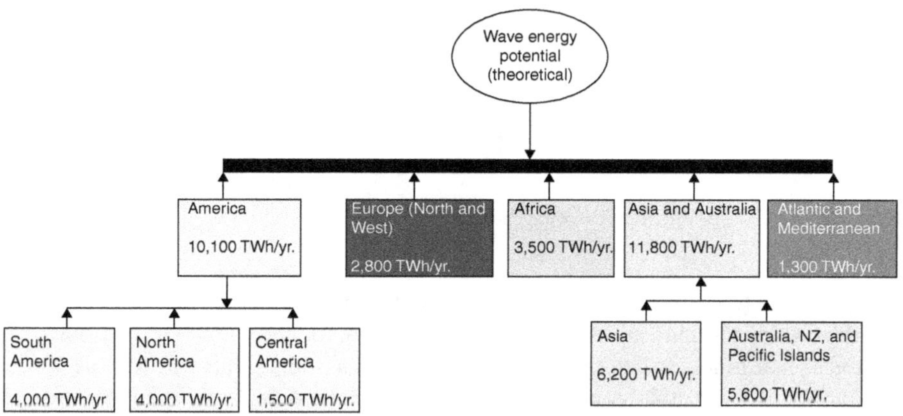

FIGURE 7.14 Global annual ocean wave energy potential.

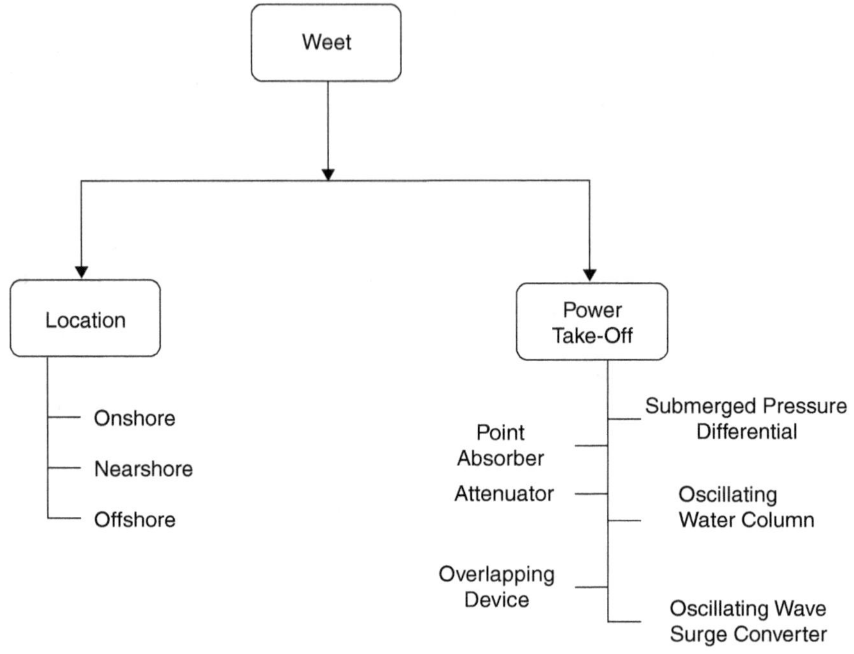

FIGURE 7.15 Classification of wave energy extract technology (WEET).

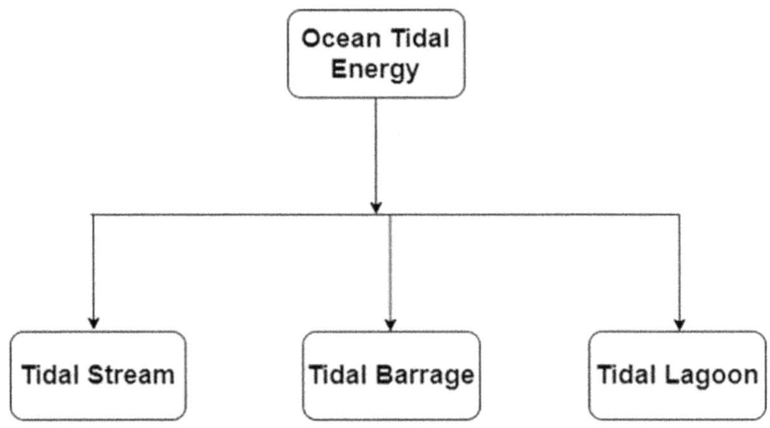

FIGURE 7.16 Classification of ocean tidal energy.

Tidal stream: In the tidal stream, a turbine is placed in fast-flowing water created by the tide. Such turbines are effective in shallow waters and allow ships to navigate around the turbines. It does not harm aquatic life as tidal generator turbine blades turn slowly. The first tidal power station was created in Strangford Lough in Northern Ireland in 2007. The tide in this strait moves at 13 feet per second across the strait. It has a severe environmental impact which has resulted in some countries including the USA not installing it.

Tidal barrage: This type of tidal generator uses a large dam known as a barrage. It is constructed across estuaries, bays, and tidal rivers. The gate of the barrage is open as the tide rises and closes at high tide resulting in a pool. Hence, it harnesses the power of the tide as a river dam harness the power of a river. Environmental impact is significant as the land is completely disrupted. Additionally, salinity of tidal barrage results in harm to animals and plants. A tidal barrage is expensive compared to a single turbine because a barrage requires more machines and construction. It also requires constant supervision. La Rance tidal power plant in France is a tidal barrage type that was built in 1966.

Tidal lagoon: It involves the construction of a lagoon. It is constructed along a natural coastline and it generates continuous power. It works as the lagoon is filling and emptying hence it has minimal environmental impact. It appears as a sea wall at low tide and submerged at high tide which allows aquatic life to swim around and inside it, depending on their sizes. Nevertheless, low energy output is obtained from tidal lagoon hence there is none in existence. However, China is constructing one dam on the Yalu River close to North Korea. A private firm is implementing one in Swansea Bay in Wales.

Tidal currents are affected by the ebb cycle and flood, but ocean currents are unidirectional. The major types of tidal current energy are depicted in Figure 7.17.

Ocean thermal energy conversion (OTEC): Sunlight falling on the ocean results in thermal energy being absorbed and stored in the upper layer. The temperature difference between warm seawater at the surface of the ocean and cold seawater to generate

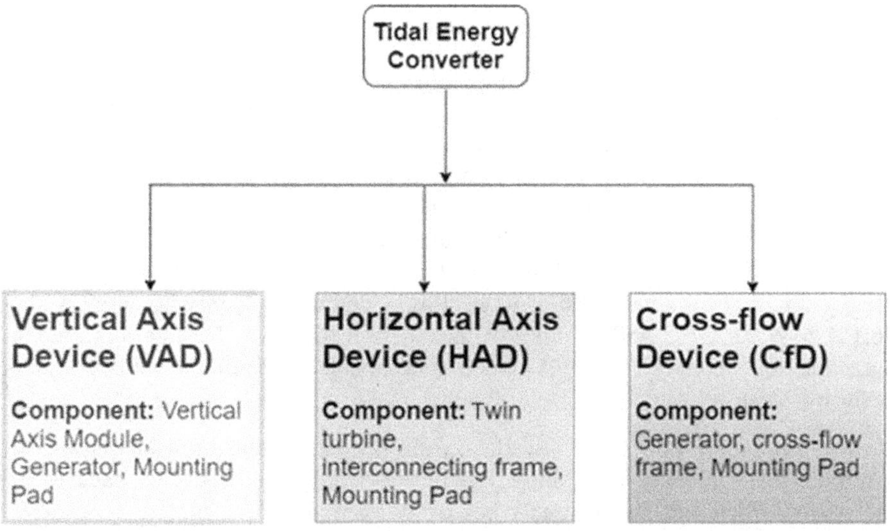

FIGURE 7.17 Classification of tidal current energy converter.

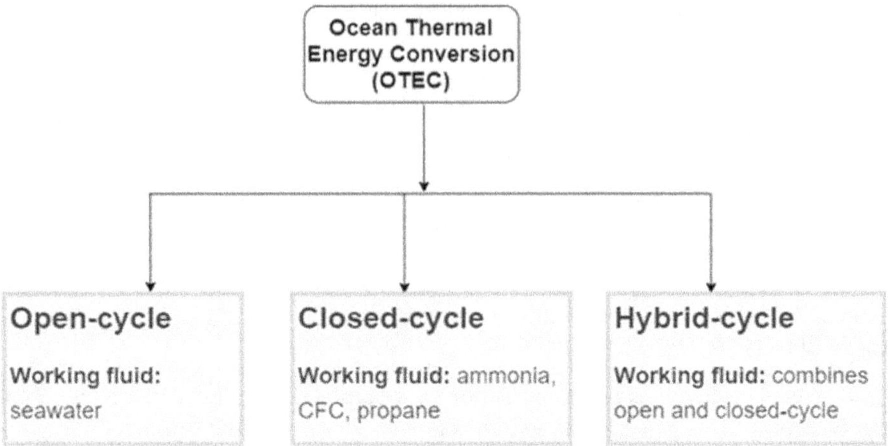

FIGURE 7.18 Classification of ocean thermal energy conversion.

electricity. This is between 800 and 1,000 m depth. Warm seawater produces working fluid for turning the turbine and cold water condenses vapor. The vapor pressure difference turns the turbine. Hybrid-cycle, closed-cycle, and open-cycle are used for generating power from ocean thermal energy as depicted in Figure 7.18.

The prospect of ocean energy is promising as it is more than over 100 years. A key challenge is the environmental and social impact deployment of ocean energy may cause. Infrastructure challenges of ocean energy are tied around the grid and supply chain. Equipment of ocean energy may corrode easily storms from the sea may damage it often.

7.4.4 Enhanced Geothermal Energy

It is the energy stored underneath the earth's crust. It resulted in the formation of the earth (20%) and the decomposition of radioactive materials (80%). Geothermal gradient causes the conduction of thermal energy from underneath to the earth as heat. The geothermal gradient is the temperature difference of the earth's crust and surface. It is used in 26 countries and geothermal heating in 70 countries. Countries such as New Zealand, Iceland, the Philippines, Kenya, Costa Rica, Indonesia, and El Salvador generate about 15% of their electricity with geothermal energy and Iceland uses geothermal for about 90% of its heating. Geothermal energy potential is about 16,127 MW for 2022 with about 24% installed in the USA. Despite this immense potential, only about 6% has been tapped. It is capable of meeting about 5% of global needs by 2050.

Geothermal power stations: It uses heat from the earth's crust to heat water to turn a turbine that generates electricity. The fluid is cooled and recycled back to heat the source. There are about three types of geothermal power stations as shown in Figure 7.19.

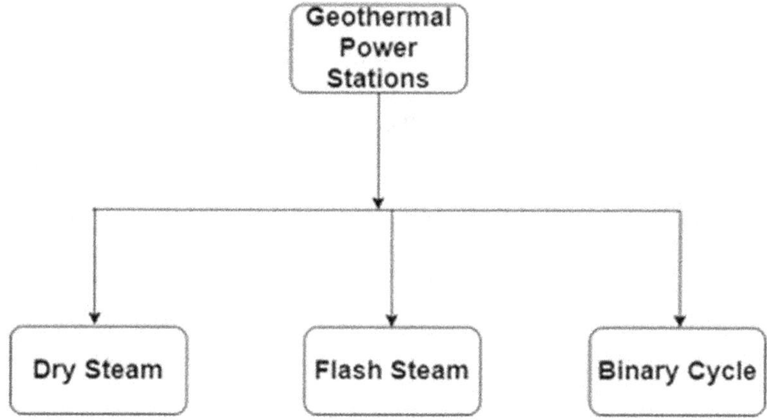

FIGURE 7.19 Classifications of geothermal power stations.

The dry steam geothermal power station is the simplest, most efficient, few in existence, and oldest type. They produce dry steam of about 150°C or more to turn a turbine that generates electricity. Thereafter, the steam generated is condensed to liquid and used to cool the water. After cooling, it is channeled for reheating and reused. It reduces every 30 years as it will need to be replenished as in the case of Geysers in California.

Binary cycle power stations are the newest innovations capable of working with a fluid temperature of 57°C or lower. It has a thermal efficiency of about 13% as it uses the Kalina and standard Organic Rankine cycle. A secondary fluid, with a lower boiling point, cools the hot geothermal water, resulting in the flash vaporization of the secondary fluid used to turn the turbines.

Flash steam power station uses high-pressure hot water in a lower-pressure tank to generate flash steam of about 180°C to turn the turbine. It is sustainable as condensed steam is injected back into the system.

7.4.5 Concentrated Solar Photovoltaics

It is a method of concentrating solar heat onto a tiny region to generate energy. Utilizing mirrors or lenses, CSP technology directs sunlight into a receiver where a primary circuit (thermal energy carrier) absorbs heat that is then used either directly or indirectly through a secondary circuit to power a turbine. CSP with a heat storage system could produce energy in overcast conditions and after sunset. CSP with thermal storage enables a much greater capacity factor and dispatchability in relation to wind and photovoltaics.

Solar energy collector is an essential part of every CSP system. The solar collector is in charge of converting solar energy into heat energy. It turns incident solar light into heat by absorbing it. Heat transfer fluid (HTF) that is passing through the collector at that point removes the heat energy. The connection between the collector and the power-producing system is made via HFT.

TABLE 7.1
Key parameters of four major CSP technologies

Technologies parameters	Parabolic trough	Linear Fresnel receiver	Solar tower	Parabolic dish
Annual efficiency	15%	8–10%	20–35%	25–30%
Hybrid mode possibility	Yes	Yes	Yes	Limited cases
Water cooling (L/MWh)	300 or dry	300 or dry	200 or Dry	None
Solar fuels	No	No	Yes	Yes
Land occupancy	Large	Large	Medium	Small
Storage possibility	Yes	Yes	Yes	Yes

Table 7.1 presents the key parameters of the four CSP technologies as compiled by [29].

All CSP plants, whether they have storage or not, are outfitted with fuel-powered backup devices to provide consistent production and the requisite capacity. Fuel burners, whether they burn fossil fuels, biogas, or solar fuels, supply energy to the heat transfer fluids or liquids. At a lesser price compared to if the facility just relied on the solar field and thermal storage, fuel-powered backups enable practically total production capacity assurance. CSP has the storage and backup capabilities that allow it to greatly improve electrical networks. Given the lower thermal storage cycle losses than other energy storage technologies, thermal storage methods used by CSP are more efficient and less expensive.

CSP technologies are not currently being used widely. Only 345 MW of built-up capacity total in California between 1985 and 1991 is still in use today. Over the past few decades, interest in CSP has increased. Since 2006, a number of new CSP facilities have become operational thanks to falling investment costs and stable energy generation costs (LCOE).

The largest CSP electricity producer is Spain. Several projects are currently in the development stage or are already under development in the USA and North Africa. Many additional power stations have been added as a result of the growing interest in CSP technologies.

Types of concentrated solar power technologies: CSPs are divided into four major categories according to the technology used to gather solar energy (collector type) as shown in Figure 7.20.

Parabolic dishes (PD): In this type, the sunlight is beamed at a focal point placed above the center of the dish. The receiver and dish move simultaneously in tracking the sunlight. Microturbines or stirling engines are the receivers of a PD. Cooling water or heat transfer liquid is not needed in PD, thereby causing a high heat-to-electricity conversion ratio. A major merit of PD is the attainment of 30% conversion efficiency. Moreover, cooling is not required, can be used in stand-alone or remote locations, is restricted to flat terrain, modular system, simplicity of fabrication, and can be easily mass produced. The demerits of PDs are lack of commercialization, complexity of

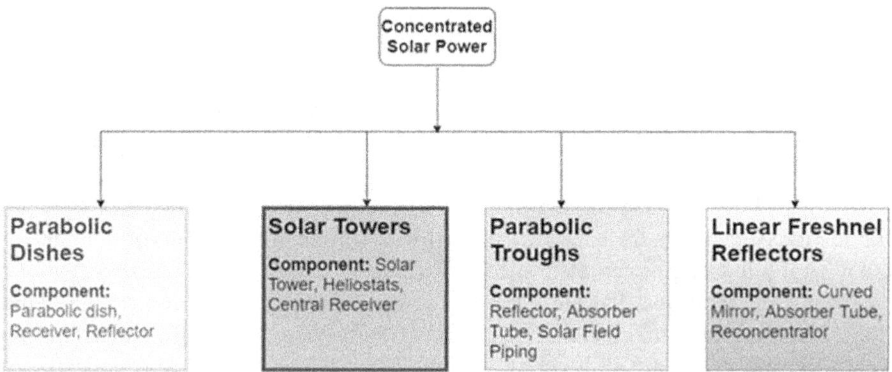

FIGURE 7.20 Key components of concentrated solar power technologies.

grid connection, and huge operational cost, among others. Europe has the largest capacity with about 10 kW, followed by USA, and about 100 MW being constructed in Australia.

Solar towers: Heliostats are used to track the sun over two different axes. Heliostats are computer-assisted mirrors. They beam the sunray on a single receiver, thereby making the concentration higher than PD, PT, and LFR. ST receiver is mounted on the central tower resulting in heat and electricity generation. ST heat transfer fluids include synthetic oil, steam, and molten salt with temperatures varying from 250° to 1,000°.

Merit includes easy deployment in hilly terrain, mid-term efficiency, and the idea for dry cooling. It has the same demerit as with other CSPs. Plans are ongoing for deployment in the USA, Spain, Germany, and Israel.

Parabolic trough (PT): Sunlight is focused using parabolic mirrors to position heat receivers on the focal line in PT technology. A unique coating is applied to the receivers to maximize energy absorption and decrease infrared re-irradiation. Receivers are enclosed in evacuated glass shells to prevent convection heat losses. Superheated steam is generated by transferring a heat-transfer liquid, such as synthetic oil or molten salt, to a steam generator. Turbine and electricity are powered by superheated steam.

About 850 MW is the total installed capacity of PTs. The majority of PT plants are found in Spain and the USA and range in size from 14 to 80 MW. The main benefits of PT include a net plant efficiency of approximately15% that has been commercially validated, modular systems, storage capacity, commercially validated investment prices, hybrid concept, lower operating costs, and considerations for land use. High thermal losses, a lack of a suitable channel for heat transmission, and lengthy pipes connecting the array to the steam generator area few of the main drawbacks of PT technology.

Linear Fresnel reflector (LFR): Like PT collectors, FR collectors use a number of lengthy, flat, slightly curved mirrors. Sun would shine on either side of a fixed receiver if these mirrors were placed at opposite angles. A few meters above the main mirror field is where the fixed receiver is placed. Each mirror line has a single-axis tracking system that is installed and customized specifically to focus the sun's rays on a fixed receiver. Unlike PT collectors, astigmatism may cause the focus line of LFR collectors to be distorted. Any missing tubes are filled in with a secondary reflector to bring the photons back into focus.

Compact linear Fresnel reflectors (CLFRs), a more current LFR technology, use two identical receivers for each row of mirrors to reduce the amount of land needed and a parabolic trough to create a specific output.

Three LFR facilities that are currently in operation have all been running since 2008. There are currently three of them. New South Wales, Australia, Murcia, Spain, and California, USA, are all home to such plants with 5, 4, and 1.4 MW capacities of these facilities, respectively.

LFR technology has several key advantages over parabolic troughs, including readily accessible materials, lower manufacturing and installation costs, the ability to generate steam directly from water, the ability to perform hybrid operations, and lower heat transmission losses. Some of the drawbacks connected with LFR are performance, investment, and operation expenses are not yet practically demonstrated, solar to electric efficiency is 8–10%, which is lower than that of PT, and combined with thermal storage is complex.

7.4.6 Artificial Photosynthesis

This method mimics nature to produce, store, and transport sustainable and efficient energy. It is the process of generating electricity by using sunlight to split water into hydrogen and oxygen. It has been employed for producing hydrocarbon fuels that can replace fossil fuels. It is an innovative way of replicating biochemical reactions known as photosynthesis. It is a reactor fueled using hydrogen energy and bioelectric transducers.

Stages of artificial photosynthesis: The first stage is harvesting chlorophyll II using sunlight. This is followed by tyrosine. Tyrosine is a catalyst used to accelerate electron transport. Third and last is the manganese complex used for transporting electrons to the chlorophyll as depicted in Figure 7.21.

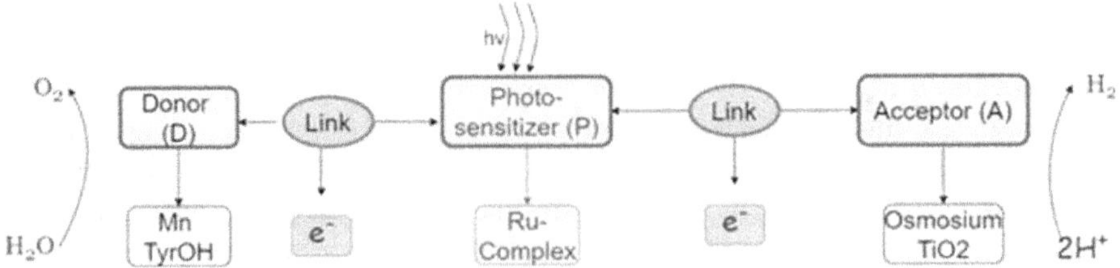

FIGURE 7.21 A schematic depiction of artificial photosynthesis.

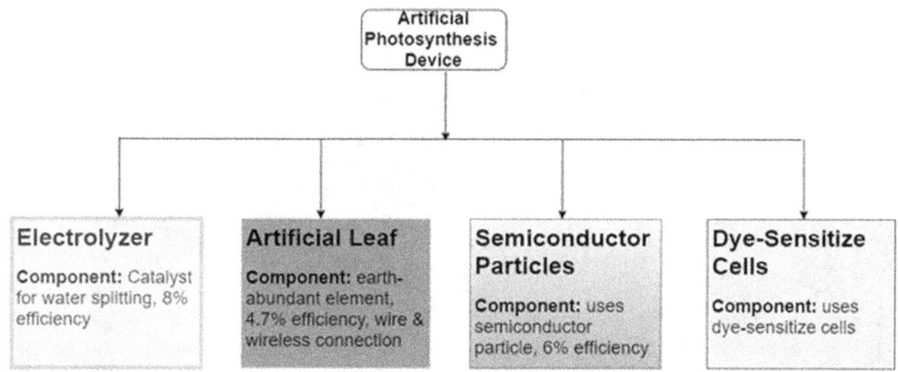

FIGURE 7.22 Schematic of artificial photosynthesis devices.

Devices used for artificial photosynthesis: Devices have been developed for generating electricity via artificial photosynthesis. They are artificial leaves, electrolyzers, dye-sensitized solar cells, and semiconductor particles as shown in Figure 7.22.

Artificial leaf: This device involves earth-abundant elements operating close to neutral pH boundary conditions. It is a solar water-splitting cell. It can either be connected using wire or wireless. An efficiency of about 2.5% has been recorded for wireless and 4.7% for wired connections.

Electrolyzer: This type employs a catalyst for the water splitting. In this case, a catalyst is deposited on the electrode's surface. An efficiency of 8% has been recorded for commercial-scale electrolyzers. The high cost of the electrode is the major impediment to this technology.

Semiconductor particles: An efficiency of about 6% has been recorded for this type. It excites electrons into higher energy gaps using semiconductor particles. Two types exist known as single and double semiconductor particle methods.

Dye-sensitized cells: This uses cells sensitized by dye. This dye could be synthetic or natural. Dye excites the electron to achieve the water-splitting process.

7.4.7 Pay As You Go Solar Energy

A new emerging technology is a technology that allows users to own their electricity based on their purchasing power. It targets low-income households in the Global South to access electricity and meet their energy needs. Different packages exist as provided by different service providers. The major package includes two to three light bulbs and a USB outlet for charging devices such as mobile phones or mobile phone power banks. The medium package includes other electronic devices such as rechargeable fans, and television. As the purchasing ability of the user increases, items such as refrigerators are included. It is a technology that is spreading across Africa and other developing regions of the world like India. It ensures the attainment

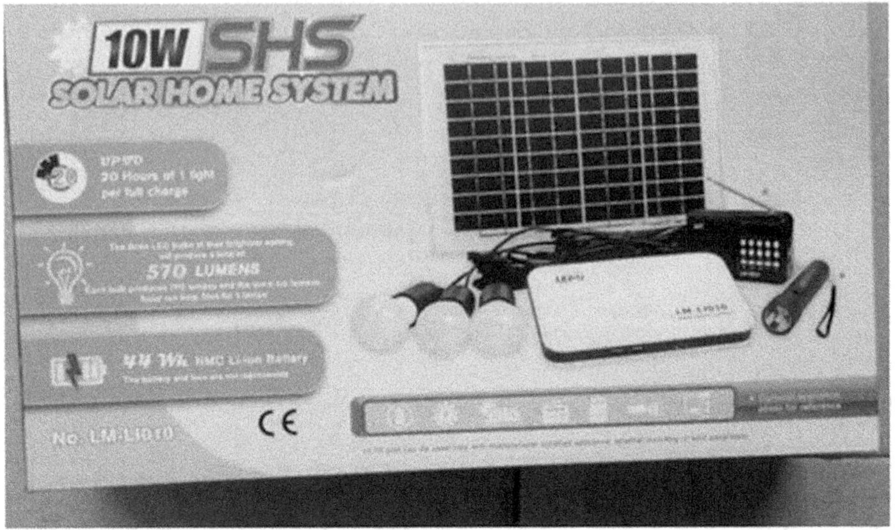

FIGURE 7.23 An example of pay as you go solar energy.

of the SDGs as it provides access to clean energy at affordable cost. A good example of pay-as-go solar energy is illustrated in Figure 7.23.

7.4.8 Cellulosic Ethanol

Ethanol is made from a range of plant materials as a feedstock and is a member of the alcohol family. This serves as the liquid gasoline for automobiles. Sugarcane and maize starch are both the principal raw materials used to make ethanol. A drawback is that the raw materials are used for food resulting in competition. In addition to these traditional techniques for making ethanol from edible substances, there are some new technologies that use grass, wood, or other plant remains or parts that are not edible. These cutting-edge methods all use plant cellulose to produce ethanol. Cellulosic ethanol is the name for the ethanol created when wood is processed.

As opposed to the technique of creating ethanol from sugarcane or cornstarch, the production of cellulosic ethanol is more difficult. Scholars at the University of California, Berkeley have calculated that cellulosic ethanol has the ability to reduce greenhouse gas emissions by 90% over the course of its life when compared to conventional gasoline made from petroleum. One study revealed that cellulosic ethanol made from particular feedstocks might be carbon zero. This indicates that a greater amount of carbon dioxide can be expelled from the environment during the entire life cycle process of cellulosic ethanol than will be released into the air.

Three generations of biofuels exist, namely first, second, and third. First-generation biofuels consist of oil, sugar, and starch, which are readily converted to ethanol, diesel, and butanol using standard conversion techniques. Lingo-cellulosic biomasses, such as farm waste and special energy crops, are used to make cellulosic ethanol, which is regarded as a second-generation biofuel. These second-generation biofuels have

TABLE 7.2
Key parameters of major technologies of cellulose energy

Major technologies	Energy harvesting method/tool	Major energy source	Drawbacks
Cellulolysis	Pre-treatment, hydrolysis and fermentation	Inedible parts or remnants of plants	Difficulty in shipment, car engine upgradation, contamination of fuel
Gasification	Fermentation and distillation		

a more involved conversion procedure. Third-generation biofuels, also known as "drop-in" fuels, are chemically and functionally comparable to gasoline. These are made from a variety of environmentally friendly materials, including cellulose, municipal refuse, and algae. Cellulosic ethanol also referred to as lingo-cellulose ethanol, can be produced in several methods. Two main methods that the scholars believe have the greatest promise for extracting cellulosic ethanol are cellulolysis and gasification.

However, research on cellulose is not yet commercialized but labeled as the next generation of energy from ethanol. Four major feedstock sources for cellulose exist. Class one is word residue, produced from wood, sawmills, furniture, and paper mills. Class two is agriculture derived from straw, maize stover, husks, and bagasse residues. Dedicated energy crops, such as herbaceous, woody, and tall grasses, make up the third class. Municipal solid wastes, consisting of paper and other cellulosic materials, are the base at the end of the chain.

The demerit of cellulosic fuel is that it is highly corrosive, absorbs atmospheric water, it is incompatible with existing car engines. Thus, it will require modification on car engines.

Cellulosic ethanol approach: Cellulosic ethanol can be achieved using a biological or thermochemical approach. The key types of cellulose energy, as adapted from Hussain, Arif, and Aslam [30], are depicted in Table 7.2 and thereafter discussed.

Thermochemical approach (gasification): This method uses partial combustion to convert carbon in feedstocks to synthetic gas. It breaks the feedstock into carbon dioxide (CO_2), carbon monoxide (CO), and hydrogen gas (H_2). These mixtures are mixed with a fermenter and Clostridium to achieve fermentation. Clostridium ingests gas produced to give ethanol and water. Desalination is used to separate ethanol from water as shown in Figure 7.24.

Biological approach (cellulolysis): Cellulolysis is a hydrolysis reaction that converts cellulose into cellodextrins, which have lesser polysaccharides, or entirely into glucose molecules. In comparison to the disintegration of other polysaccharides, cellulolysis is comparatively challenging given that cellulose molecules bind tightly to one another. The detailed process of the biological approach is depicted in Figure 7.25.

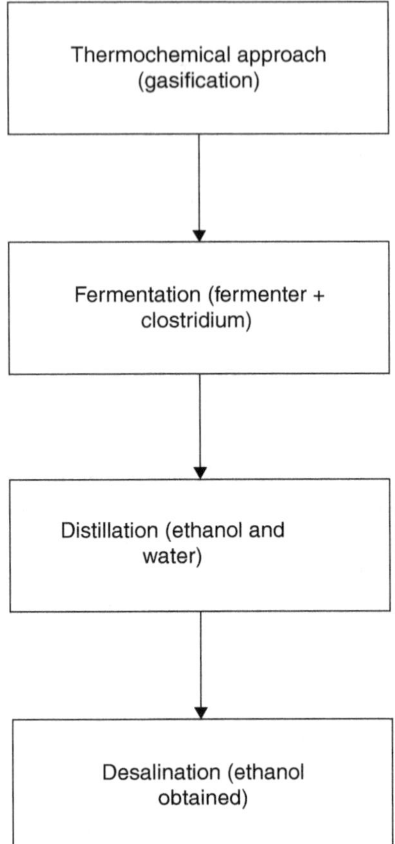

FIGURE 7.24 Steps for gasification of cellulose.

7.4.9 Solar Panels that Work at Night (Solar Thermo Electric Generator)

This technology improves on the limitation of solar energy. Conventional solar panels can only produce electricity during active sunlight (daytime solar irradiance). However, infrared solar panels work in the dark and during rainfall. It works by harvesting the energy of radiative cooling of a solar cell at night.

How it works: It operates using radiative cooling as depicted in Figure 7.26.

Heat is emitted when the earth cools from the setting of the sun, thereby creating a temperature difference between the solar panel surface and the air. A thermoelectric generator is mounted to the solar panel to harvest this radiative cooling energy.

Assawaworrarit, Omair, and Fan [31] achieved 50 mW/m^2 and 100 mV open circuit voltage nighttime electricity at night. This was achieved by using a thermoelectric generator that harvests energy using solar cell's temperature difference and ambient surroundings.

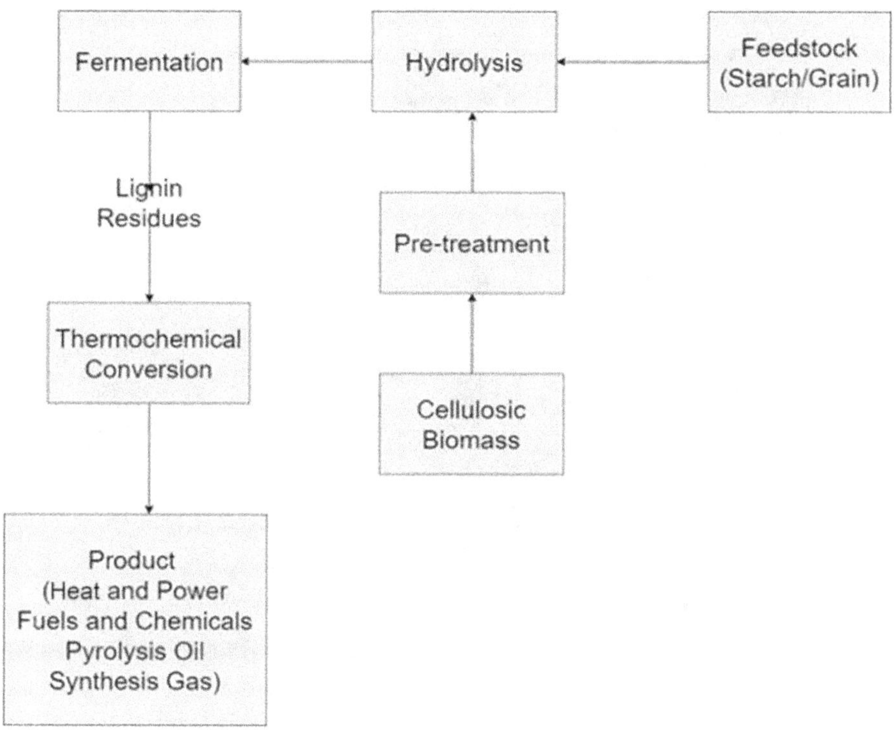

FIGURE 7.25 Schematic of cellulolysis process.

7.4.10 Hydrogen Energy

Energy generated from other sources can be transported and stored using hydrogen. Several processes can be used to generate hydrogen fuel. Electrolysis and natural gas reforming, a thermal procedure, are the two most popular techniques. About 1 kg of hydrogen can produce 33 kWh of electricity. About 2.3 gallons of water to create 1 kg of hydrogen which equals 423 standard cubic feet. Hydrogen can replace gasoline in internal combustion engines. Although it is less efficient and produces tailpipe emissions. One kg of hydrogen produces 2.205 pounds of energy which is equivalent to 2.8 kilogram or 6.2 pounds of gasoline. Also, 14.128 liters is equivalent to 1 kg of hydrogen. About 6 kg of pressurized tanks is used for hydrogen-powered cars for 400 miles. About 70 kWh of energy is produced from every kilogram of hydrogen for a car. Demerits of hydrogen include high cost, high flammability, and difficulty encountered in transporting in longer distances. Companies such as Toyota are now investing in hydrogen because it behaves like gasoline but quieter and doesn't pollute the environment.

Research is ongoing for green hydrogen integration with current sustainable energy sources. There is currently interest in using green hydrogen as a carbon-neutral solution integrating renewable energy sources. The danger of an uneven supply or demand in the electric grid due to the intermittent behavior of renewable energy resources is

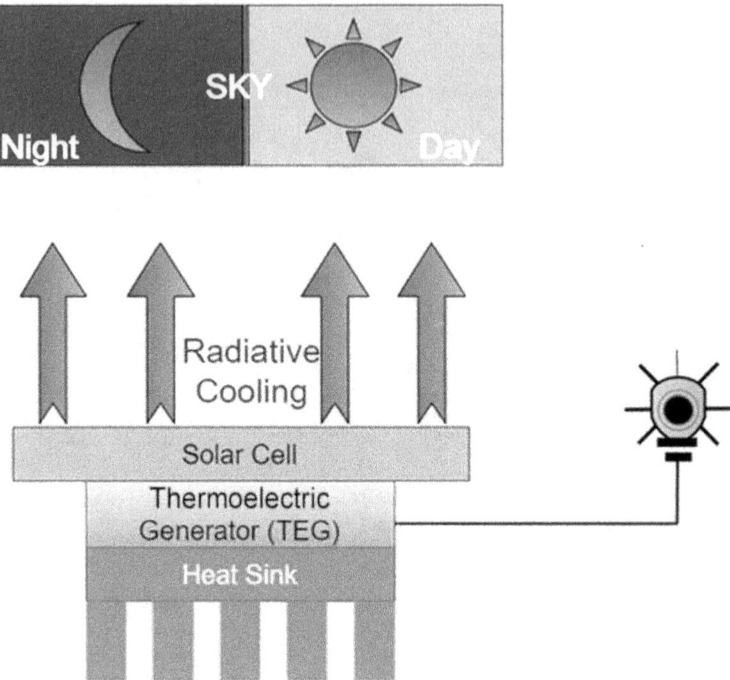

FIGURE 7.26 A schematic of solar in the dark.

one of the biggest problems with the direct integration of solar and wind energy into the grid. Oversupplying the grid with energy during high wind or daylight hours can be just as harmful as not having access to enough energy to meet demand. A device for maintaining energy storage could be made of green hydrogen.

Types of hydrogen: Apart from green hydrogen, other types of hydrogen exist along the hydrogen color spectrum. They are green, gray, blue, pink, black and brown, turquoise, yellow, and white hydrogen. Figure 7.27 depicts the classification based on the method and types of hydrogen energy.

Grey hydrogen: It is produced using steam methane reforming from methane or natural gas. It is the most common form of hydrogen production. It is the least renewable form of hydrogen. It is low cost and used for refining oil and fertilizers production. It costs 2 USD/kg in the USA, and 5–5 USD/kg in Asia, Australia, and Europe. However, the recent increase in the cost of natural gas meant that the price has gone up more than green hydrogen. It accounts for the 830 million tonnes of carbon dioxide produced from hydrogen.

Blue hydrogen: This is produced using steam reforming. The process involves using natural gas to heat water to steam resulting in hydrogen and carbon dioxide. Blue hydrogen emits a bit of greenhouse gases, hence it is called low-carbon hydrogen. However, it uses carbon capture and storage to capture and store this carbon to

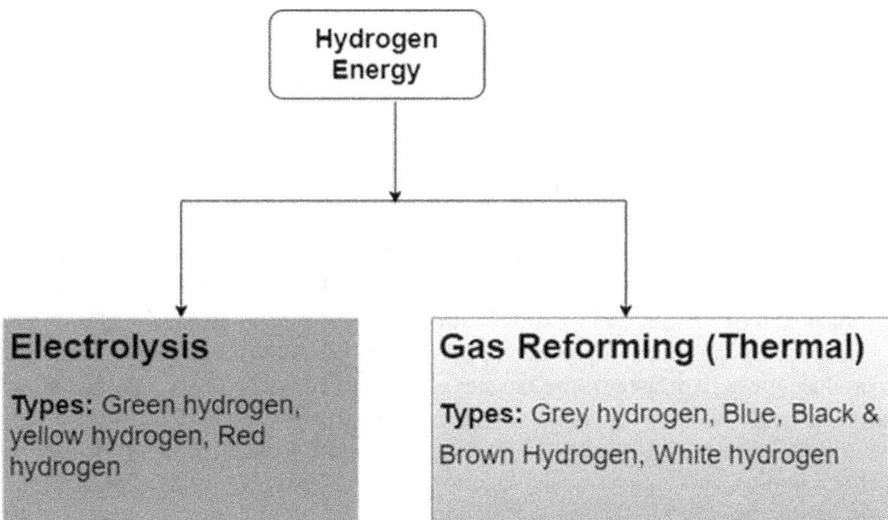

FIGURE 7.27 Classification of method and types of hydrogen energy.

improve the hydrogen produced. Carbon capture and storage are used for reducing carbon emissions to prevent global warming. It captures the carbon dioxide, transports it, and stores it underground.

Black and brown hydrogen: This is the opposite of green hydrogen as it damages the environment. It is produced from gasification using fossil fuels. Australia and Japan are producing an improved brown coal to hydrogen. A liquefied brown hydrogen is produced using Australian brown coal and is shipped to Japan for low-emission usage.

White hydrogen: It occurs naturally. It is produced from fracking of geologically occurring hydrogen. It still lacks the technology to exploit and explore it. Hydrogen in its natural and pure elemental form, is primarily found in subterranean deposits and liberated during hydraulic fracturing. In very minute amounts, it can also be found in the air. White hydrogen is a colorless, odorless, and tasteless gas made up of two hydrogen atoms bound together. It is the lightest chemical element and the first element in the periodic table. Even though it is a naturally occurring element, it is not frequently employed as an energy source as it is challenging to isolate and extract it in its purest form. The natural form of H_2 present in the air and subsurface deposits liberated during fracking is called white hydrogen. It is the lightest chemical element.

A variety of distinct sources produce the white hydrogen that can be found on earth. They are serpentinization, a reaction between ultrabasic rocks and water, is one of the sources of natural H_2. Deep inside the earth's crust and mantle, there is H_2 that is degassing. Water's interaction with recently exposed rock surfaces is known as weathering. A collision of water with reducing substances found in the earth's organic material involves the breakdown of hydroxyl ions in crystalline structures and natural radiolysis of water's biological function.

Many subterranean sources have been found in reservoirs in places like France, Mali, the USA, and more than a dozen more nations. Unfortunately, they would all be difficult to reach, and it would be difficult to gauge the reservoirs' potential to decide whether the effort would be worthwhile.

Ocean floor emanations are another source of white hydrogen, but it would be difficult to use them, thus scientists have turned to other sources like water electrolysis.

Green hydrogen: This is a renewable energy that does not pollute the environment. It is the energy obtained from water splitting using electrolysis. The final product is hydrogen and oxygen. It requires a vast amount of energy for the water-splitting process. It uses wind or solar for the electrolysis. Many methods of electrolysis were used to produce green hydrogen from seawater as a supply of water. These four techniques are solid oxide electrolysis, alkaline electrolysis, proton-exchange membrane electrolysis, and direct electrolysis of saltwater. The green hydrogen produced is used for fueling airships, cars, spaceships, and among others. About one-third of global green hydrogen is produced by China with an output of 20 million tons. It costs 10 to 15 USD/kg. The cost of producing 1 kg is about 6 USD. A price of 2 USD/kg is needed to achieve commercial viability.

Yellow hydrogen: It is similar to green hydrogen but the energy for electrolysis is through solar energy.

Red hydrogen: It is similar to green hydrogen but energy for electrolysis is from nuclear energy. It is also called purple or pink hydrogen. The high temperature of the nuclear reactor is used for producing steam for electrolysis or in fossil gas-based steam methane reforming.

Turquoise hydrogen: It is produced using methane pyrolysis. It is a low-emission hydrogen that is the newest among the hydrogen types.

Uses of hydrogen: It is used for different applications including transportation, heating, green ammonia production, natural gas industry among others. Hydrogen is used for internal combustion engines or fuel cells. It is used for heavy-duty vehicles and cars. Some companies in Europe are now developing heavy-duty vehicles to run on hydrogen (especially green hydrogen). Airbus is also leading in the design of aircraft powered by hydrogen. However, Airbus believes full implementation will be accomplished by 2050 by commercial aircraft powered by hydrogen that will be ready by 2035. Hydrogen can be used for heating and cooking. It is now proposed as a replacement for heating in the United Kingdom. Nevertheless, some believe the cost of electrical heating will still be better than hydrogen for a while. Green hydrogen is used for the production of fertilizers by producing a major constituent of the fertilizer known as green ammonia. Hydrogen council estimates that gray ammonia will be replaced by green ammonia because of the cost difference.

Global development of hydrogen: The hydrogen market grossed 900 million USD as of 2020. Fitch Solution gives a 10% increment by 2030. Goldman Sachs estimates it to gross 1 trillion USD by 2050. More projects have sprung up since 2020. About

136 projects representing 121 GW costing 500 billion USD as of 2021 are now in various stages of development across the world.

Asia is the current leader in hydrogen. China represents one-third of hydrogen production globally. Sinopec in China is targeting an output of about 500,000 tonnes of hydrogen production before 2025. Hydrogen electrolyzer was used for the winter Olympics of 2022 in China to fuel cars used at the games. Onshore wind was used to provide energy for the electrolyzer used for generating the green hydrogen. In India, government-owned Oil India Limited launched a pure green hydrogen plant in Duliajan (eastern Assam) in 2022. The aim of the plant is to reduce the cost of manufacturing, storing, and transporting green hydrogen in India. The Union Ministry of Power intends to generate green hydrogen of 5 million tonnes capacity by 2030. Also, Adani and Reliance Industries have commenced plans to use hydrogen in India. Reliance industries target 400,000 tonnes of hydrogen-powered by 4 GW solar energy for water splitting. A 30-billion hydrogen facility powered by 25 GW of solar and wind energy will commence and be completed in 2028 and 2038, respectively in Oman. An industrial-scale solar-powered green hydrogen plant was launched in 2021 in Dubai, United Arab Emirates. Japan plans to make the country a hydrogen society by 2030. About 135 hydrogen-powered stations currently operate in Japan with 1,000 more being developed. South Korea has also implemented the Clean Hydrogen Energy Portfolio Standard (CHPS).

In Africa, a 10 GW project has been launched in Mauritania on green hydrogen tagged NOUR project. The second project handled by the country is called the EMAN project. It is an 18 GW wind energy and 12 GW solar energy powered project to produce green ammonia worth 1.7 million tons per annum. The green ammonia produced is exported and used in the country. The project is in partnership with CWP of Australia. Namibia is following the same step as the country partnered with Germany and Holland for funding of green hydrogen. Egypt, Morocco, and Tunisia are other African countries working assiduously to implement green hydrogen.

The hydrogen Strategy for a Climate-Neutral Europe was introduced by the EU in July 2020 with the objective of becoming carbon-neutral by incorporating hydrogen into EU plans. In 2021, a resolution endorsing this tactic was approved by the European Parliament. The initial phase, which runs from 2020 to 2024, aims to decarbonize all current hydrogen output. Phase 2 (2024–2030) will incorporate green hydrogen into the energy system. In the third period (2030–2050), hydrogen will be employed extensively in the decarbonization process. By 2050, according to Goldman Sachs, hydrogen will make up 15% of the energy balance in the EU.

Germany, Austria, France, the Netherlands, Belgium, and Luxembourg are the six EU members who have asked that legislation be put in place to support financing for hydrogen. German investments totaling €9 billion have already been made to build 5 GW of hydrogen power by 2030. A lot of the members have intentions of acquiring hydrogen from abroad, particularly from North Africa. These initiatives could step up the manufacture of hydrogen. However, they have also come under fire for allegedly attempting to export the required adjustments to Europe. Beginning in 2021, all newly manufactured gas turbines within the EU must be prepared to consume a hydrogen-natural gas mixture. Thirty businesses unveiled a groundbreaking initiative to supply hydrogen located in Spain in February 2021. By the end of the 10 years, the project hopes to have built 67 GW of electrolysis capacity and 93 GW of solar power. Portugal revealed intentions to build the first solar-powered hydrogen production facility by April 2023. Galp Energia, an energy company with its headquarters in Lisbon, disclosed intentions to build an electrolyzer to power its refinery by 2025.

The British government released its "Ten Point Plan for a Green Industrial Revolution" in 2021. This paper outlined policy recommendations and encompassed investments to produce 5 GW of low-carbon hydrogen by 2030. The plans call for collaboration with businesses to carry out the required testing so that by 2023, all homes connected to the gas grid will be able to blend up to 20% hydrogen into their gas supply. Despite this, a BEIS consultation in 2022 indicated that grid blending would only play a "limited and temporary" role because less natural gas was anticipated to be used. A plan to use Scottish offshore wind energy to power converted oil and gas rigs into a "green hydrogen hub" that would feed fuel to nearby distilleries first surfaced in March 2021. Equinor revealed intentions to triple UK hydrogen production in June 2021. A proposal to integrate green hydrogen into the grid with a 200-meter wind turbine that powers an electrolyzer to generate gas for about 300 homes was unveiled by National Grid in March 2022. Energy companies like ERM, Source Energie, and RWE expressed interest in producing green hydrogen using floating wind generators in the Celtic Sea in the early 2020s. In 2025, Vattenfall plans to produce renewable hydrogen off Aberdeen using an offshore wind test.

President George W. Bush announced a $1.2 billion proposal for the creation of hydrogen fuel cell cars during his 2003 State of the Union Address and dubbed it "freedom fuel." Barack Obama stopped funding this proposal in 2009. The Federal Infrastructure Investment and Jobs Act, which became law in November 2021, allocates $9.5 billion to green hydrogen projects. The US Department of Energy (DOE) intended to launch the nation's first hydrogen network test in Texas in 2021. The department had earlier tried to implement the Hydrogen Energy California project. Texas is the largest domestic producer of hydrogen and already has a hydrogen pipeline network, making it a crucial component of green hydrogen projects nationwide.

7.4.11 Batteries (Emerging Technology)

Renewable energy is fast gaining attention. This is because of the environmental and sustainability issues surrounding conventional energy sources. Solar and wind energy are the top renewable energy sources. Their interest is connected to global availability, improvement in affordability, sustainability, and environmental mild. However, the issue of reliability acts as a major limitation for solar and other renewable energy. The battery is one of the key storage devices for solar and other sources. Different research is ongoing to improve the cost, efficiency, and sustainability of batteries and ways of making them environmentally friendly.

A battery is a chemical device that stores electrical energy in the form of chemicals and uses electrochemical reactions. It converts the stored chemical energy into DC electric energy.

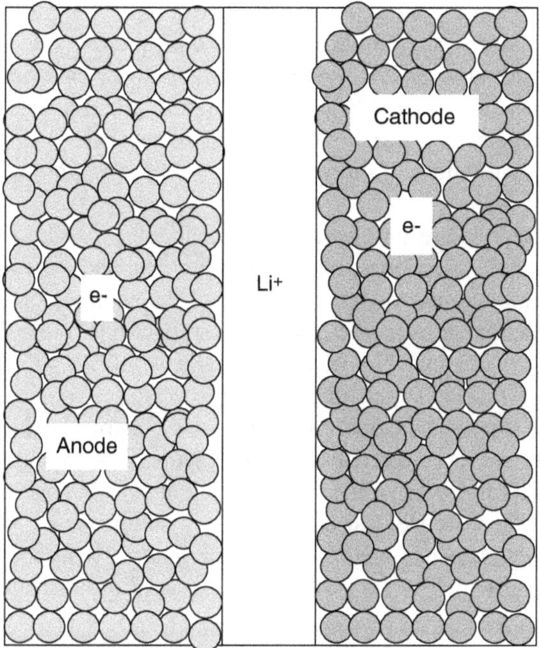

FIGURE 7.28 Schematic of battery composition.

Figure 7.28 illustrates a section of a battery. The component of the battery is encased in a steel container. Inside the enclosure is the positive terminal called cathode separated by separator from the negative terminal by the anode and all immersed in electrolyte. Silvery matte rings made of electrolyte, graphite, and manganese dioxide make up the cathode. The zinc paste inside the separator serves as the anode. To avoid a short circuit, the separator maintains the electrodes apart.

Alessandro Volta, an Italian physicist, invented the first battery in 1800. The electrochemical reaction in a battery involves the transfer of electrons from one material to another (called electrodes) through an electric current. Different types of batteries exist, which is determined by the anode and cathode materials. There are various techniques for manufacturing batteries. Mines, materials, electrodes, cells, packs, vehicles, recycling. There are different anode and cathode materials. However, the most common anode material is graphite. Currently, there is a shift toward silicon and lithium-metal batteries. However, lithium-ion batteries are gaining interest. The most common cathode material is Nickel Manganese cobalt (NMC) but there is a shift toward nickel-rich cathodes. The key performance metrics for batteries include C-rate, capacity measured in Ah, and energy measured in Wh. Batteries have different form factors namely cylindrical, prismatic, and pouch. Different battery technologies exist based on the chemistry of their cathode and anode materials. This includes lead acid batteries, nickel–cadmium, lithium-ion batteries, nickel-metal hydride, lithium metal (lithium-transition metal oxide), lithium-sulfur, and metal-air batteries.

Batteries are storage devices used for storing and charging power. Emerging technologies of batteries are lithium-metal anodes, silicon, solid-state electrolytes, lithium–sulfur (Li-S), advanced Li-ion designs, redox flow batteries (RFBs), sodium-ion (Na-ion), Zn-ion, Zn-air, Zn–Br batteries. Battery plays a crucial role in the durability and performance of equipment. The battery comes in various sizes and types depending on the equipment. Batteries are often designed for specific categories of equipment.

Lithium-ion batteries are the most used in technology such as electric cars, and electronic devices such as phones, tablets, and laptops, and have a huge potential in advancing autonomous microsystems because of their unique characteristic, safety, reliability, and energy storage performance compared to other kind of batteries. Companies that are interested in energy storage would largely benefit from this work, especially those interested in lithium-ion batteries. For example, electronic device companies, medical device companies the electric car industry (e.g. Tesla, Coda, BMW, Volkswagen, Mercedes, Toyota, Hyundai Automotive). Our work can be meaningful to the general public (not just scientists). It gives the general public a more optimal battery to use. For example, by having a better battery, automotive companies can give their customers a better experience driving their cars with longer driving range, faster battery charge time, and the ease of not having to connect the plug into the car.

7.4.12 Small Modular Reactor

It is a modern nuclear power plant with producing capacities of up to 300 MW(e) per unit, or approximately one-third of those of conventional nuclear power plants. Small modular reactor (SMR) units are currently operational in Russia, India, and China. Moreover, more than 80different modular reactor designs and incomplete demonstration projects globally. Rolls-Royce contributes to the United Kingdom's continued development of cutting-edge methods to combat the worldwide danger posed by climate change through its SMR. NuScale, USA is leading the SMR drive in the USA. In 2008, Bill Gates founded TerraPower to scale SMR. Also, Kairo engineers are developing a molten salt reactor. SMR project takes about 5 years to complete but large reactors may reach 12 years before completion. Reactors can be single modules or a combination of different units. It can be easily fabricated in a factory but has a budget of about 3 billion US dollars. It is termed Modular because each unit can fit into another and is easily scaled to serve the electricity requirement of places. SMR can easily connect to existing electricity networks. This makes it easy for it to replace dilapidating electricity stations. SMRs vary from the more prevalent nuclear power reactors of current in a couple of keyways: Initially design simplicity facilitates their integrated strategy to manufacturing.

Small modular reactors are easier to create because most of the work is done off-site. Second, SMRs have many passive or inherent safety measures built into their design. This indicates that, in the case that power fails to the plant, operator involvement or a separate power source is not required to shut down the reactor and keep cooling to get rid of the core's leftover heat. Nuclear power plants have a tiny geographic footprint in contrast to other sources, such as hydropower, wind, and solar power. In comparison to the massive reactor sites currently in use around the world, tiny modular reactors will need an even lower footprint.

For disaster planning, traditional power plants have a 16-km radius, with a larger quarantine zone of up to 80 km to safeguard water and food supplies. Evaluation of socioeconomic, environmental, and engineering requirements and costs are all part of the process of selecting a location for a nuclear reactor. SMRs provide more site options than giant reactors for a number of factors related to their construction and limited geographic footprint. SMRs use very little gasoline and only need to refill every 2 years or so. Additionally, they do not need freshly created reactor fuels, such as fuels that are accident-tolerant and have high levels of safety. Passive safety systems that enable reactors to "walk away safe" allow SMRs to operate on conventional reactor fuel.

7.5 SPINTRONICS

A newer technique known as spintronics, or "spin transport electronics," also referred to as magnetoelectronics, makes use of the electron's intrinsic spin and magnetic moment in addition to its basic electronic charge. It is a research investigation of the electron's intrinsic spin and related magnetic in solid-state devices, along with the electron's core electronic charge.

With implications for the effectiveness of data storage and transfer, spintronics essentially varies from conventional electronics in that electron spins are used as an extra level of freedom besides to charge state. In the fields of quantum computing and neuromorphic computing, spintronic systems are of special interest and are often realized in Heusler alloys and dilute magnetic semiconductors (DMS).

History of spintronics: In order for SPICE developers and later circuit and system designers to utilize the physics of spin transport for the investigation of spintronics "beyond CMOS computing," a generalized circuit theory for spintronic integrated circuits has been suggested.

Spin-dependent electron transport processes in solid-state devices were discovered in the 1980s, and this led to the development of spintronics. This includes Baibich et al.'s [32] discovered giant magnetoresistance as well as Johnson's [33] observation of spin-polarized electron injection from a ferromagnetic metal to a nonmagnetic metal (1988). Meservey and Tedrow's groundbreaking ferromagnet/superconductor tunneling tests and Julliere's early studies of magnetic tunnel junctions from the 1970s are credited with giving rise to spintronics. The speculative development of a spin field-effect transistor by Datta and Das in 1990 and the electric dipole spin resonance by Rashba in 1960 marked the beginning of the use of semiconductors for spintronics.

The theory of spintronics: The angular momentum of the electron's spin, which is distinct from the angular momentum resulting from its orbital velocity, is intrinsic. The spin-statistics theory indicates that the electron behaves as a fermion because the magnitude of the electron's spin projection along any axis is 1/2 h. The spin has a

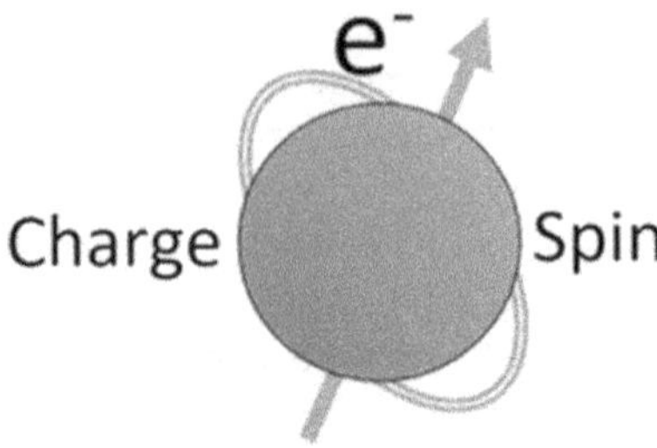

FIGURE 7.29 Schematic of spintronic.

corresponding magnetic moment, just like orbital rotational momentum. Figure 7.29 depicts the schematic of the spintronic.

Application: Mass-storage systems make use of spintronic technology. It is used to fit a lot of data into a tiny space; for example, a single-sided 3.5-inch disc can fit about 1 TB of data at 1.5 Gbit/mm^2 or one trillion bits per square inch. There is a lot of research being done on non-volatile spin-logic devices to allow scaling. Devices with spin-transfer, torque-based logic that process information by using spins and magnets have been recommended. The exploratory road plan for the ITRS includes these devices. Applications for logic-in-memory are already under development. Bhatti et al. [34] reviewed it.

7.5.1 Clean Energy and 3D Printing

3D printing is also changing manufacturing, energy, and other sectors of the economy of developed and developing nations. 3D printing is improving the efficiency of wind and tidal turbines. It is now being used for turbine retrofits. The energy sector is benefitting from the usage of 3D printing due to the following: the ability of 3D printing to produce custom-made parts, repair and replicating complex components or machines, design flexibility of 3D printing, and scaled and improved component testing and evaluation.

In a hydroturbine, the feasibility of using 3D printing to produce Axial Flow Hydropower Turbine Housing, Runner, and Draft Tube has been studied by [35]. Moreover, researchers worked on rapid manufacturing by polyjet technology of customized turbines for renewable energy generation [36]. To integrate design and production for a tiny Pelton turbine (less than 5 kW), the study suggests a novel approach. The application of rapid product development (RPD) for a brand-new, highly tailored, ultra-compact hydropower turbine was covered. For the wind turbine, a PC-ABS filament was used to 3D print a 1.7-m wind turbine blade with a 178 mm chord, and pitch of 6° modular aerodynamics [37]. The wind turbine strain change was also measured in a wind tunnel of 11 m/s and change in flapwise bending moment. Other studies exist that used 3D printing for wind turbines.

For tidal energy, the thermomechanical performance of a printed tidal turbine utilizing Digimat-fused AM's filament production technique. The thermoplastic polymers polyamide 12 (PA12), ABS, polyamide 6 (PA6), and polyetherimide (PEI) are the filament. These thermoplastic polymers are reinforced with carbon and glass fillers in the form of fibers and beads (CF/GF and CB/GB). In comparison to PA6-CB/CF, it is demonstrated that the blade printed with PEI-CB/CF exhibits outstanding mechanical performance with minimal mechanical deflection and warpage. Additionally, for 3D printing, fiber-shaped fillers perform better than bead-shaped ones. Overall, this study has shown the viability and promise of 3D printing as a fantastic possibility for the production of small blades.

Majority of the applications of 3D printing in the solar energy sector are made in concentrated solar panel heating and molten salt. It is also observed that SLA 3D printers and sintering are widely used.

7.6 FUTURE OF EMERGING TECHNOLOGIES

Things are evolving to meet the growing population and human needs. Technologies are evolving to make life easy. Technologies aim at improving people, planet, and places as they affect life (human and non-human). Energy, food, climate, water, and sanitation will be priorities for researchers working on developing new technologies or improving on existing technologies.

7.7 SUMMARY

This chapter discussed emerging technologies with emphasis on the latest and all-encompassing. Technologies such as AI, spintronics, self-healing materials, energy from Wi-Fi, self-fertilizing crops, robotics, biotechnology, nanotechnology, gene therapy, 3D printing, stem cell therapy, and multi-use rockets were discussed. Fossil fuels are being improved with solar, wind, hydro, and geothermal positions to play key roles in clean and sustainable energy. Also, emerging clean energy technologies were examined in relation to their applications, merits, demerits, and opportunities. Some of them already found wide applications. 3D printing continues to blaze the trail as it is found in various applications. The application of 3D in building, clean energy, and others were discussed to expose the vast potential of this technology.

REFERENCES

1. Srinivasan, R., 2008. Sources, characteristics and effects of emerging technologies: Research opportunities in innovation. *Industrial Marketing Management, 37*(6), pp.633–640.
2. Rotolo, D., Hicks, D. and Martin, B.R., 2015. What is an emerging technology? *Research Policy*, *44*(10), pp.1827–1843.
3. Zhu, M., Yu, J., Li, Z. and Ding, B., 2022. Self-healing fibrous membranes. *Angewandte Chemie*, *134*(41), e202208949.
4. Gaddes, D., Jung, H., Pena-Francesch, A., Dion, G., Tadigadapa, S., Dressick, W.J. and Demirel, M.C., 2016. Self-healing textile: Enzyme encapsulated layer-by-layer structural proteins. *ACS Applied Materials & Interfaces*, *8*(31), pp.20371–20378.

5. Lee, J. and Park, Y., 2023. Self-healing properties of fibers constructed from mushroom-derived chitinous polymers. *ACS Sustainable Chemistry & Engineering*, *11*(7), pp.2959–2967.
6. Alsaba, Y., Rahim, S.K.A. and Leow, C.Y., 2018. Beamforming in wireless energy harvesting communications systems: A survey. *IEEE Communications Surveys & Tutorials*, *20*(2), pp.1329–1360.
7. Ding, Z., Perlaza, S.M., Esnaola, I. and Poor, H.V., 2014. Power allocation strategies in energy harvesting wireless cooperative networks. *IEEE Transactions on Wireless Communications*, *13*(2), pp.846–860.
8. Chen, Y., Zhao, N. and Alouini, M.S., 2017. Wireless energy harvesting using signals from multiple fading channels. *IEEE Transactions on Communications*, *65*(11), pp.5027–5039.
9. Schutter, G. De, Lesage, K., Mechtcherine, V., Nerella, V.N., Habert, G. and Agusti-Juan, I., 2018. Vision of 3D printing with concrete – technical, economic and environmental potentials. *Cement and Concrete Research*, *112*, pp.25–36.
10. Romani, A., Rognoli, V. and Levi, M., 2021. Design, materials, and extrusion-based additive manufacturing in circular economy contexts: From waste to new products. *Sustainability*, *13*(13), p.7269.
11. Schweiker, M., Endres, E., Gosslar, J., Hack, N., Hildebrand, L., Creutz, M., … Roswag-Klinge, E., 2021. Ten questions concerning the potential of digital production and new technologies for contemporary earthen constructions. *Building and Environment*, *206*, 108240.
12. Guerrero-Lemus, R. and Martínez-Duart, J.M., 2012. *Renewable Energies and* CO_2 (2012 Edition, (pp.9–31). Springer.
13. IEA, 2022. *Wind Power Generation in the Net Zero Scenario, 2010–2030*. IEA. www.iea.org/data-and-statistics/charts/wind-power-generation-in-the-net-zero-scenario-2010–2030-2, IEA. Licence: CC BY 4.0.
14. Siebenhüner, B., Arnold, M., Eisenack, K. and Jacob, K.H., 2013. *Long-Term Governance for Social–Ecological Change*. Routledge.
15. Jia, Y., Wei, J., Wang, K., Cao, A., Shu, Q., Gui, X., … Ma, B., 2008. Nanotube–silicon heterojunction solar cells. *Advanced Materials*, *20*(23), pp.4594–4598.
16. Barber, J., 2009. Photosynthetic energy conversion: Natural and artificial. *Chemical Society Reviews*, *38*(1), pp.185–196.
17. Frouin, R. and Pinker, R.T., 1995. Estimating photosynthetically active radiation (PAR) at the earth's surface from satellite observations. *Remote Sensing of Environment*, *51*(1), pp.98–107.
18. Vasarevicius, D. and Martavicius, R., 2011. Solar irradiance model for solar electric panels and solar thermal collectors in Lithuania. *Elektronikair Elektrotechnika*, *108*(2), pp.3–6.
19. Chendo, M., 2002. Factors militating against the growth of the solars-PV industry in Nigeria and their removal. *Nigerian Journal of Renewable Energy*, *10*(1&2), pp.151–158.
20. Agbo, S. and Oparaku, O., 2006. Positive and future prospects of solar water heating in Nigeria. *Pacific Journal of Science and Technology*, *7*(2), pp.191–198.
21. Sahu, B.K., 2015. A study on global solar PV energy developments and policies with special focus on the top ten solar PV power producing countries. *Renewable and Sustainable Energy Reviews*, *43*, pp.621–634.
22. IRENA and IGA, 2023. *Global Geothermal Market and Technology Assessment*. International Renewable Energy Agency and International Geothermal Association.

23. Nakada, S., Saygin, D. and Gielen, G., 2014. *Global Bioenergy Supply and Demand Projections for the Year 2030.* International Renewable Energy Agency. Available at www.irena.org/-/media/Files/IRENA/Agency/Publication/2014/IRENA_REmap_2030_Biomass_paper_2014.pdf
24. Takahashi, P. and Trenka, A., 1996. *Ocean Thermal Energy Conversion.* Wiley.
25. Crus, J., Ed., 2008. *Ocean Wave Energy: Current Status and Future Perspectives.* Springer.
26. Pelc, R. and Fujita, R.M., 2002. Renewable energy from the ocean. *Marine Policy, 26,* pp.471–479.
27. OES, 2014. *Implementing Agreement on Ocean Energy Systems.* OES Annual Report.
28. Falcão, A.F.D.O., 2010. Wave energy utilization: A review of the technologies. *Renewable and Sustainable Energy Reviews, 14*(3), pp.899–918.
29. Barlev, D., Vidu, R. and Stroeve, P., 2011. Innovation in concentrated solar power. *Solar Energy Materials and Solar Cells, 95*(10), pp.2703–2725.
30. Hussain, A., Arif, S.M. and Aslam, M., 2017. Emerging renewable and sustainable energy technologies: State of the art. *Renewable and Sustainable Energy Reviews, 71,* pp.12–28.
31. Assawaworrarit, S., Omair, Z., Fan, S. 2022. Nighttime electric power generation at a density of 50 mW/m2 via radiative cooling of a photovoltaic cell. *Applied Physics Letters*, 120(14).
32. Baibich, M.N., Broto, J.M., Fert, A., Nguyen Van Dau, F.N., Petroff, F., Etienne, P., … Chazelas, J., 1988. Giant magnetoresistance of (001)Fe/(001)Cr magnetic superlattices. *Physical Review Letters, 61*(21), pp.2472–2475.
33. Johnson, M., 2002. Spin injection in metals and semiconductors. *Semiconductor Science and Technology, 17*(4), p.298.
34. Bhatti, S., Sbiaa, R., Hirohata, A., Ohno, H., Fukami, S. and Piramanayagam, S.N., 2017. Spintronics based random access memory: A review. *Materials Today, 20*(9), pp.530–548.
35. Chesser, P., Love, L., Witt, A.M. and Roos, P., 2020. *Feasibility of Using Additive Manufacturing to Produce Axial Flow Hydropower Turbine Housing, Runner, and Draft Tube.* Oak Ridge National Laboratory (ORNL).
36. Udroiu, R., Nedelcu, A. and Deaky, B., 2011. Rapid manufacturing by polyjet technology of customized turbines for renewable energy generation. *Environmental Engineering and Management Journal, 10*(9), pp.1387–1394.
37. Abdelrahman, A. and Johnson, D.A., 2014. Development of a wind turbine test rig and rotor for trailing edge flap investigation: Static flap angles case. *Journal of Physics: Conference Series, 524*(1), 012059.

8 Corrosion and Additive Manufacturing

8.1 BACKGROUND OF THREE-DIMENSIONAL PRINTING

Three-dimensional (3D) printing is the creation of an object through "printing". The object is created using a computer-aided design software, which is then converted to a format for 3D printing. The most common format is .stl. Elements called resins or filaments are used as the feedstock for the 3D printing. Thin layers of material – such as liquid or powdered plastic, metal, or cement – must first be laid down. Next, the layers must be fused in tandem.

However, in general, 3D printing involves the following modeling, slicing, printing, and post-processing as illustrated in Figure 8.1.

3D printing is an emerging technology that has found application in various sector of the economy. It has been deployed in automotive, warfare, space, building and construction, among others. Designers use 3D printers to rapidly produce product models and prototypes, but they are also increasingly being used to produce finished goods. The products created with 3D printers include toys, furnishings, tools, tripods, gift and novelty items, wax castings for jewelry, and shoe designs.

It is adaptable, flexible, can be customized and becoming affordable. A 3D printer costs less than 500 USD. However, the cost goes higher depending on the type and the application. It uses filament or resins depending on the printer type and application. Different materials have been used for the filament. These range from metals of different types, polymers, ceramics, with the list growing by the day.

Fused deposition modeling (FDM) is the most prevalent method of 3D printing that primarily uses polymer filaments. Inkjet printing, contour crafting, stereolithography (SLA), direct energy deposition (DED), and laminated object manufacturing (LOM) are the primary methods of additive manufacturing of powders by selective laser sintering (SLS), selective laser melting (SLM), or liquid binding in three-dimensional printing (3DP), as well as inkjet printing, contour crafting, SLA, DED, and LOM. In the consumer sector, there are now two main 3D printing technologies: FDM vat polymerization. SLA, masked stereolithography (MSLA), and digital light processing make up vat polymerization (DLP).

The extruder, print bed, nozzle, and filament used by the printer are the four main components of a 3D printer. Other components are the moving parts, and touch screen. The printing bed is the platform on which models are printed, and it is frequently

DOI: 10.1201/9781003441151-8

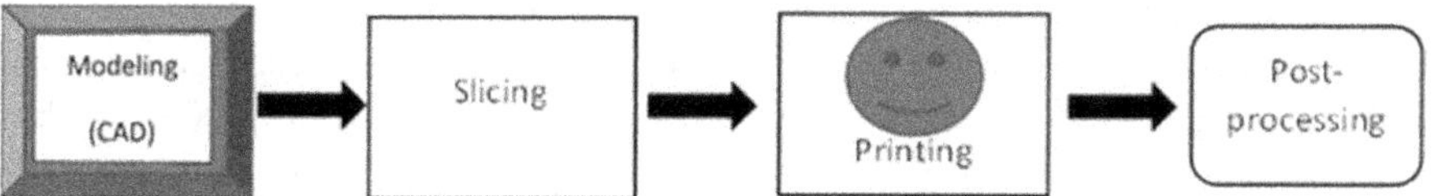

FIGURE 8.1 Schematic of 3D printing steps.

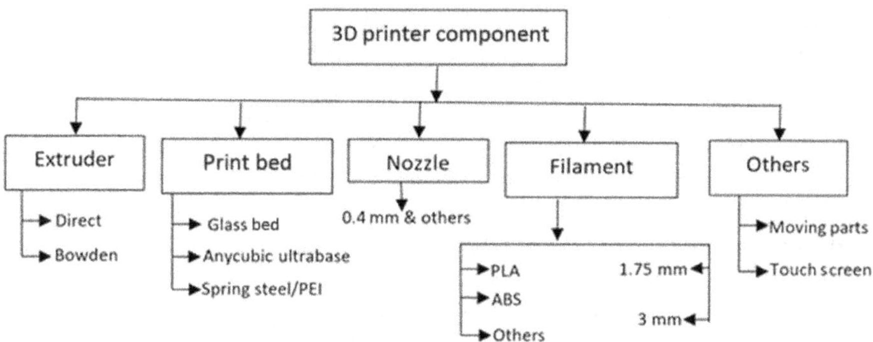

FIGURE 8.2 Key components of 3D printer.

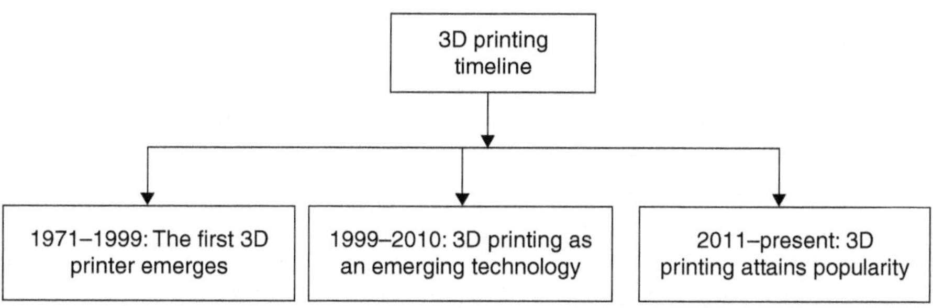

FIGURE 8.3 Timeline of 3D printing technology.

heated to assist layers adhere to each other tightly. Figure 8.2 presents the summary of the key components of a 3D printer.

The history of 3D printing dates to over 80 years ago. Nevertheless, it gained prominence a few years back. The timeline for 3D printing is illustrated in Figure 8.3.

The first patent for a liquid metal recorder was issued in the 1970s, but the concept is far older. In 1945, a futuristic short story by Murray Leinster called "Things Pass By" described the process of feeding "magnetronic plastics – the stuff they make houses and ships of nowadays – into this moving arm. It scans drawings using photo-cells and produces drawings in the air. Between 1971 and 1999, the first 3D printer was developed. This is divided into two eras, viz the pre 1990s and the 1990s.

The pre 1990s saw the discovery of 3D printing technology. In the 1990s, the main 3D printer manufacturers and CAD tools emerged. The first recorded variations of 3D printing may be traced back to the early 1980s in Japan. Hideo Kodama was looking for a technique to design a quick prototyping system in 1981.In 1984, French engineers invented SLA. Charles Hull began studying SLA in 1986, and the first SLA-1 machine was built in 1988. DTM Inc. built the first SLS machine, which was later purchased by 3D System in 1988. In this era, the first EOS Stereos system was released in 1990. In 1992, Stratasys received the FDM patent. Solid scape began operations in 1993. In 1995, the MIT granted Z Corporation an exclusive license. The next era was from 1999 to 2010. Despite public fears that the millennium bug would cause computer systems to crash and digital Armageddon, 3D printing showed considerable promise in a variety of businesses. During this era, RepRap and MakerBot 3D printers were made and gained prominence facilitated by the need to reduce cost to meet consumer needs. RepRap 3D printer built a 3D printer in 2004. Spectrum Z510 was introduced by Z Corp in 2005. It was the world's first high-definition color 3D printer. The year 2006 saw the emergence of an open-source project (Reprap). The first 3D printed prosthetic limb was created in 2008. FDM patents are released into the public domain, and Sculpteo was founded in 2009. The current era of 3D printing started around 2011 to date. 3D printing is now a well-established technique. Consumer interest in industrial platforms expanded. Some believe additive manufacturing will eventually replace traditional CNC and milling manufacturing.

8.2 CONCEPT OF 3D PRINTING

3D printing is a term used for a group of technologies primarily in industry, referred to as additive manufacturing. These technologies create items by layering materials on top of each other, based on digital data. These components are then further processed and combined into products such as jet engine components, hearing aids, medical implants, and a variety of other complicated industrial parts. 3D printing is a supplement to, not a replacement for, traditional manufacturing processes. For decades, industry has employed 3D printing to develop prototypes for product design and production planning. With the technologies having advanced, they are being employed to create end-use parts and products. Additive manufacturing is particularly beneficial for producing unique parts and small batches of complex things at a lower cost than traditional manufacturing, which frequently necessitates time-consuming and costly equipment preparation.

3D printing has two important characteristics that make it a "green" technology [1]. Unlike traditional production techniques such as injection molding, casting, stamping, and cutting, many 3D printing systems produce relatively little waste. Second, digital designs may be printed on-site by 3D printers in homes, stores, and community centers, minimizing the requirement for things to be transported to end consumers. 3D printing has attracted the attention of the public, engineers, and environmental visionaries during the last decade [2]. It has been heralded as a manufacturing revolution as well as a possibility for significant environmental improvement

[3]. However, quantitative measurement of 3D printing's environmental performance is lacking. Much of it focuses solely on energy consumed during production, ignoring the effects of raw material production, product consumption, and waste management.

8.3 CORROSION AND IMPACT ON 3D PRINTING (FILAMENTS AND TECHNIQUES)

Although most materials corrode, selecting the appropriate material reduces or prevents corrosion. 3D printing is not exempted from corrosion as conventional materials are used for 3D printing operation. 3D printers are also being developed to print a variety of materials, including plastics, metals, composites, and many more. When it comes to industrial 3D printing, there are many materials to choose from. These materials have distinct characteristics, strengths, and shortcomings. Furthermore, essential elements such as material type, texture, pricing, and others must be considered to avoid 3D printing errors. It can be tough to select the best material for a particular job. Attempts have been made to classify the materials used for 3D printing into metals and alloys, ceramics, polymers and composites, concrete, hybrid, among others as illustrated in Figure 8.4.

i. *Polymer and composite 3D printing:* Polymer 3D printing is a new technique that is witnessing greater application in industry, particularly in the medical field. Polymer is a natural or manmade substance made of long molecules called macromolecules that are multiples of smaller chemical units known as monomers. Key characteristics and qualities of many engineering polymers include corrosion resistance, light weight yet comparatively stiff and strong, resilience, toughness, lack of conductivity (heat and electrical), color, transparency, processing, and low cost [4, 5].

Some examples of polymer 3D printed materials include acrylonitrile butadiene styrene (ABS), acrylic-based (Stratasys: MED 610), methacrylic acid, polycarbonate (PC), polyether ether ketone (PEEK), polyethylene terephthalate glycol (PETG), polylactic acid (PLA), and polyamide 12 (Nylon). Different 3D printing processes use polymer as the 3D printing material. These include FDM (ABS, PETG, PC, PEEK, PLA), multijet fusion (nylon), polyjet (acrylic-based) [6], and SLA (epoxy-based and methacrylic acid) [7].

ii. *Metal and alloy 3D printing*: Metals and metal alloys can now be 3D printed in increasing numbers. Most important metals and alloys are aluminum, nickel, titanium, Inconel, copper, bronze, cobalt, nickel, cobalt-chrome, precious metals (gold, silver, platinum), and steels (tooling, maraging, and stainless).

Stainless steel is perfect for objects that will come into touch with corrosive substances, water, or steam due to its superior corrosion resistance. Bronze is used to make fittings, vases, and more aesthetically pleasing objects like marine propellers and pump impellers. Jewelry can be created using gold as the printing material. Nickel can be used to create coins and even components for turbine engines. For metal products, especially those that need to be lightweight, like airplane parts, aluminum is perfect. Medical implants (such as hip joints) and other solid fixtures and things can be made from titanium to produce parts that are incredibly precise and strong.

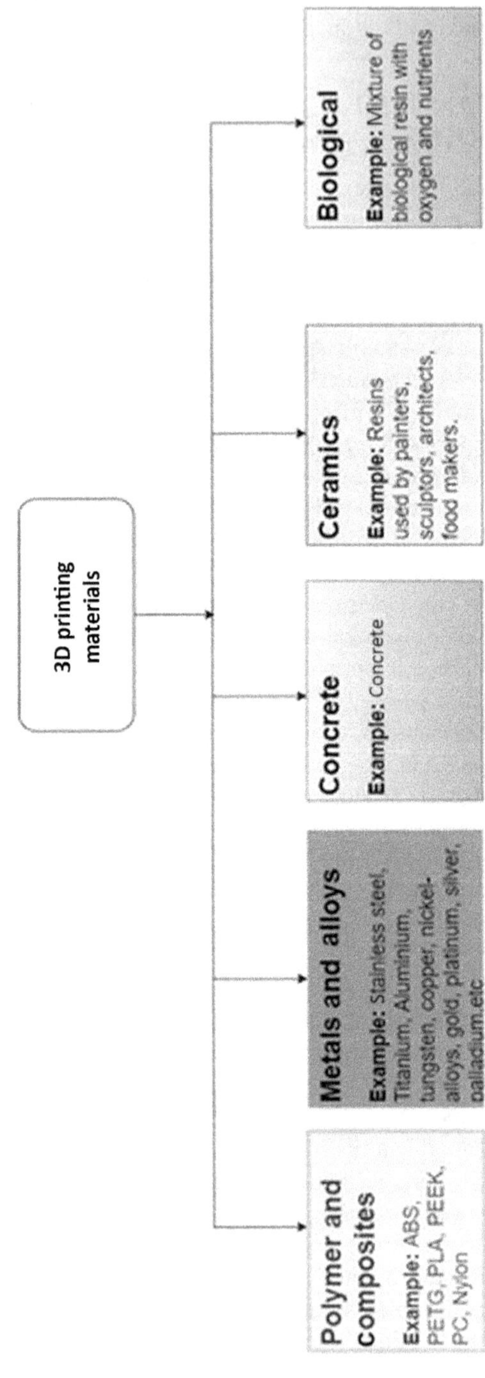

FIGURE 8.4 Classification of 3D printing materials.

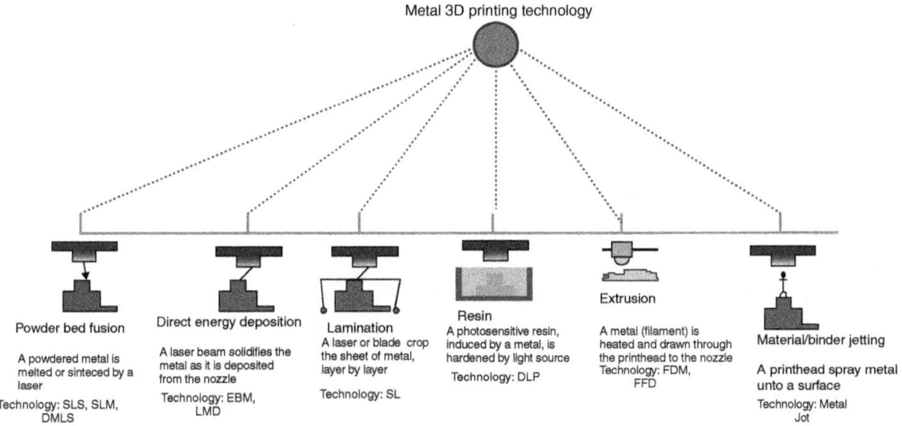

FIGURE 8.5 Summary of metal 3D printing technology.

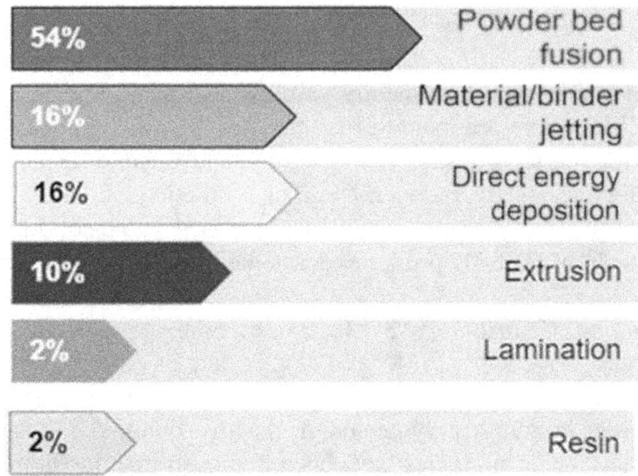

FIGURE 8.6 Distribution of metal 3D printer for 2019.

Powder, wire, and filament are the most frequent forms of metal and alloys used for 3D printing. Metal 3D printing resin and metal sheets for lamination-based 3D printers are also available. Figure 8.5 presents the summary of metal 3D printing technologies adapted from Cherdo [8].

3D printer models per technology in existence are depicted in Figure 8.6. It can be seen that powder bed fusion dominates the 3D printer market.

iii. *Ceramics and 3D printing:* Ceramic resin is a photopolymer that contains silica. The photopolymer network burns out during firing, resulting in a genuine ceramic item. Fired parts have unique features. 3D printing ceramics has inspired a new generation of painters, sculptors, and architects who are drawn to the ornate and intricate shapes that would be impossible or too time-consuming to create using

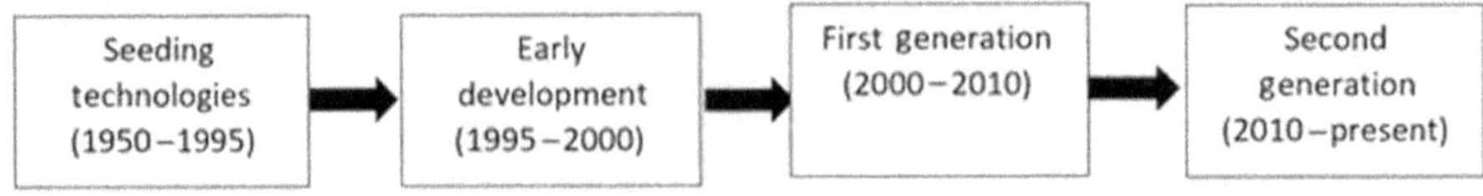

FIGURE 8.7 Trends of concrete 3D printing technologies.

TABLE 8.1
A comparison of parameters for concrete 3D printing

S/N	Parameter	Concrete printing	Contour crafting	D-shape
1.	Printing process	Extruding	Extruding	Spreading
2.	Nature of material	Mixed 3D printing concrete	Cementitious or mortar	Powder or chemical agent
3.	Speed of printing	High	Low	Medium
4.	Dimension	Large scale	Large scale	Medium size
5.	Resolution	4–6 mm	Smooth	Approx. 13 mm

conventional clay methods. It is mostly employed in technologies like SLA, DLP, multijet, and CLIP. There are many different types of resins that can be used in 3D printing, including castable resins, strong resins, and flexible resins, among others. UV-curing liquid plastics are known as resin or synthetic resin. SLA 3D printers use these materials. The photosensitive resin can be turned into a solid model by curing it with a laser or light. The 3D printer exposes the liquid resin at the required spots, curing the resin.

iv. *Concrete and 3D printing:* "3D Concrete" refers to concrete extrusion technologies, whereas "ARCS,""LSAM," and "FC" relate to other subgroups. The most common 3D concrete printing procedures are extrusion (concrete/cement, wax, foam, polymers), powder bonding (polymer bond, reactive bond, sintering), and additive welding. A variety of technologies have been demonstrated, including building and construction component fabrication on-site and off-site, using industrial robots, gantry systems, and tethered autonomous vehicles.

The history of concrete 3D printing spans several decades starting from 1950 to present as depicted in Figure 8.7.

The autonomous robotic construction system (ARCS) is a 20-by-40-foot environmentally friendly concrete printer that can construct a 1,490-square-foot home in 36 hours. The system can construct houses, businesses, roads, and bridges. ARCS can handle projects ranging in size from 500 square feet to over one million square feet.

S-Squared 3D Printers Inc. is a Long Island, New York-based 3D printer manufacturer and retailer. The company, which began operations in 2014, manufactures 3D printers for hobbyists, libraries, and STEM programs.

Table 8.1 presentsthe parameters of the various concrete 3D printing process as adapted from Zhang et al. [9].

v. *Biological and 3D printing:* The 3D printing material used in biological or medical field is different from other applications. Biological mixture comprising of oxygen, cells, nutrients and others is used for biomedical 3D printing. This is discussed in detail under biomedical 3D printing applications.

8.4 CORROSION IMPACT ON BIOMEDICAL 3D PRINTING APPLICATION

Biomedical is a subfield of medicine that integrates biological and physiological concepts into clinical practice. Biomedical research explores the impact of medications and medical methods on the biological systems of organisms. A good example of biomedical devices includes infusion pumps, artificial organs, pacemakers, dialysis machines, cochlear implants, the heart–lung machine, implants, artificial limbs, corrective lenses, ocular prosthetics, facial prosthetics, dental implants, and somato prosthetics. The growth of biocompatible prostheses, various diagnostic and therapeutic medical devices ranging from clinical equipment to micro-implants, widely used imaging devices like MRIs and EEGs, regenerative tissue growth, pharmaceutical drugs, and therapeutic devices are some notable biomedical engineering applications.

Metallic biomaterials exposed to corrosion lose both structural stability and surface functionality. It quickens wear, fretting wear, and fatigue; conversely, such damage quickens corrosion.

Corrosion in human body: Between 40% and 60% of the bulk of the human body is made up of water [10]. Practically, the extracellular and intracellular fluids make up both main fluid compartments that make up the overall amount of bodily water. Both pH and temperature have a significant impact on how materials corrode. Temperature of bodily fluids is 37°C under normal circumstances. Regarding corrosion, this can be viewed as a consistent temperature for the course of an implant's life. The pH, dissolved oxygen, and chloride levels of bodily fluids are among the crucial factors that affect how quickly metal implants corrode. Metal corrosion in living organisms can be viewed as an electrochemical reaction.

When metallic implants corrode, three things can happen: (i) electrical currents may impact cellular behavior; (ii) the corrosion process may change the chemical environment; and (iii) metal ions may modify cellular metabolism. Systemic and distant impacts are one of the problems caused by the entry of corrosion products into the body. Mild corrosion frequently causes symptoms as well, which can range from localized discomfort at the site of the corroded area to intense pain, reddening, and swelling across the entire broad area around the device.

Application of 3D printing in biomedical: The 3D printing process used in the biomedical field is known as 3D bioprinting. In 3D bioprinting, biological procedures such as merging cells or biomaterials to produce an object with tissue qualities are printed using 3D technology. Compared to non-biomedical 3D printing, it involves the layer-by-layer deposition of biological substance made up of biolinks as opposed to filament or resin. Then, the biomedical industry may exploit these tissue-like structures for a variety of purposes.

The process of 3D bioprinting is divided into various steps, viz pre-bioprinting, bioprinting, and post-processing as depicted in Figure 8.8.

i. *Pre-bioprinting:* The pre-bioprinting is the process of creating a 3D model of the object to be produced. It can be executed using 3D scanner, by editing a freeware or downloading a created object from group, and by modelling it. Biopsy is needed for the preliminary process. Tomographic remodeling is made for the layer-to-layer operation. A biological substance is used as the printing raw material. Prior to usage, the biological feedstock is multiplied, and some cells isolated to enrich them with nutrients and oxygen.
ii. *Bioprinting:* After the pre-bioprinting stage, the actual printing of the organ or tissue happens.

The printing material is a biological mix of cells screened, isolated, and enriched with nutrient and oxygen. For tissue printing, an incubator is needed for the pre-printing. The biological mix of the cells is layered on each other using a biocompatible scaffolding.

iii. *Post-processing:* This is done after the bioprinting operation, to "clean up" the produced object for use. Post-process operation depends on the object produced, the type of printer, type of finishing or location of usage, among others. FDM printer does not need a lot of post-processing operation. Post-processing is demanding because of the organic materials present. Stability of the organic material used is vital in the same way as transplanted organ needs a stable environment after transplant.

Future of 3D biomedical: Technology of 3D bioprinting is changing human life including the medical field. 3D bioprinting is employed in a variety of medical applications. It is used in prosthetics, organ, and tissue printing. However, some of the bioprinted organs like human heart are not transplanted into humans but used for disease prevention research, visualization, for medical, and drug testing, among others. A heart can be 3D-printed in 4–5 hours.

3D bioprinting is revolutionizing medicine by reducing the need for animal testing, because of advancements in 3D printing applications in medicine. Cost, flexibility, availability and compatibility of prosthetic are being improved via 3D printing. Drugs and other medical consumables are made on demand and are custom-built with the evolution of 3D printing. In fact, surgery is becoming easier and scalable with advent of 3D bioprinting. The technology offers hope and promise in the biomedical field.

Pre-bioprinting → Bioprinting → Post-processing

FIGURE 8.8 The process of 3D bioprinting.

8.5 CORROSION IMPACT ON AUTOMOTIVE, AVIATION AND AEROSPACE IN 3D PRINTING APPLICATION

The car manufacturing sector is under pressure to increase fuel efficiency because vehicle greenhouse gas emission rules are becoming more stringent around the world. Reducing vehicular mass and creating energy-efficient cars, which lessen environmental effect, are essential in accomplishing this. In the design and construction of modern vehicles, weight reduction has emerged as one of the most crucial factors to take into account, whereas electric vehicles demand a different strategy. Corrosion is a crucial factor to consider when choosing and using materials, and all the aforementioned factors have a substantial impact. Any time a car's metal surface gets into touch with moist air, atmospheric corrosion may result. Steel surfaces can slowly corrode from a thin moisture film that forms beneath non-zero humidity circumstances. When components degrade because of oxidation, car corrosion occurs. Rusting, a form of chemistry that happens when iron reacts with air and water, is typically what starts this procedure. Rust does not always suggest that corrosion is prevalent. Corrosion is simply another name for rust. Rust can weaken exhaust system mounts, corrode the trunk area, and chip away at the metal of the engine space, leading to air leakage and leaving the trunk susceptible to water spills.

Magnesium is widely used in automobiles and is the lightest engineering metal for structural uses. It has been used in instrument panel beams and steering wheels. As magnesium has the lowest (least noble) electrochemical potential of all the metals used in automotive structures, its weak corrosion protection is inherent. Mg has a porous and unprotective surface oxide coating compared to the similarly reactive metal Al, which does not offer passivation against corrosion like the passivation seen on Al alloys. The most common magnesium alloys for the automotive industry continue to be AZ91D (officially Mg–9Al–1Zn (0.4Mn), wt%) and AM60 (officially Mg–6Al–1Mn, wt%). However, there has been little advancement made in the accessibility of corrosion-resistant commercial magnesium alloys. When combined with virtually all other automotive alloys, Mg alloys are particularly vulnerable to galvanic corrosion due to its reactivity and the previously stated low electrochemical potential. One way to stop Mg alloy corrosion is by using a barrier coating, but despite their efficacy, these coatings tend to lead to adverse anode/cathode area ratios at coating defect sites. The use of aluminum in the automobile industry has grown over the past few decades, with most original equipment makers usually starting with the use of aluminum hoods. This is due in part to aluminum's excellent corrosion resistance on par with steel.

When joining carbon fiber materials with metals, corrosion problems may develop because of the addition of carbon fiber reinforcement. Carbon strands used in CFRPs are highly noble electrochemically and electrically conductive. Consequently, galvanic corrosion can occur when a metal or alloy is improperly joined to CFRP (so that there is an electrical link). The problem gets worse when a fastener, bolt, or nut is used because a sizable portion of CFRP is linked to tiny metallic components. Due to the high cathode to anode surface area ratio in these conditions, the rate of galvanic corrosion may be accelerated.

3D printing can be used to solve key challenges associated with electric vehicles. The four stages of electric vehicle production and application of 3D printing are discussed here. In the automotive sector, 3D printing has developed, moving from being mainly a prototyping technology to becoming a widely used production tool. According to recent reports, the global 3D printing industry will grow by 14.4% by 2027 and be worth $26.7 billion.

Design stage of electric vehicles and 3D application: This includes use of light-weight for the body parts, chassis configuration, fire safety, and electric safety. Electric vehicles cover maximum distance that the car can cover before the batteries are recharged. Thus, the weight of the car is considered in the design of electric car. A reduction in the components of electric car while considering the strength is key in the design process of electric car. 3D printing is used for testing, optimization, and metal 3D printing of electric cars parts. A metal 3D printer (binder jetting 3D printing) is employed for the metal printing of parts that is 50% lighter than the conventional method. As an example, General Motors increased the strength of the back seat bracket by 20% while reducing the weight by 40% using the Autodesk Generative Design tool.

Electric motors, not engines, are used to propel electric cars. Therefore, to make the vehicle lightweight, the transmission and battery pack must be positioned and optimized to fit in as little space as feasible. Hence, choosing the chassis designs requires using small-scale prototypes. To rapidly print scale models and evaluate the options, 3D printing techniques such as FDM or fused filament fabrication (FFF) are used. Batteries must be shielded from heat and fire, and because batteries produce heat during the discharge cycle, the enclosure must effectively dissipate that heat. Thus, it is necessary to produce enclosures with vent openings for good ventilation. 3D printing can be used to create intricate patterns. 3D printing of flame-retardant and flame-resistant fabrics includes material such as ULTEM 9085.

Production stage of electric vehicles and 3D printing: This includes tooling for metal and sheet metal parts, tooling for die casting, tooling for plastic parts, jigs, and fixtures. Electric motor housings, suspension mounting brackets, and steering knuckles are examples of parts that are typically manufactured using a metal casting, forging, or machining process. Later, the components are CNC machined with specialist equipment to finish the surface and drill the fastening holes. To get a precise shape with fewer tool passes and processes, specialized cutting tools are required. As an illustration, center drilling and chamfering can be merged using a unique tool that performs both tasks simultaneously. These unique tool bases can be created using metal 3D printing, and cutting tool components made of cubic boron nitride or carbide can then be attached to them. A good example is Kennametal's 3D-printed metal cutting tool for making electric car motor housing.

Stamping die tools are required for the mass manufacturing of sheet metal components. Traditionally, five-axis machining centers are used to form metal dies that can withstand thousands of stamping operations. However, FDM 3D-printed tools made of ULTEM 9085, which have been shown to resist pressures up to 70 MPa and last 300 pressing cycles, can be used for experimental purposes and form validation. Metal dies with complicated geometries and cooling channels are required for the pressure die casting of metal components. Inserts enable the creation of such complicated shapes. Recently, businesses have started looking into 3D printing dies to create conformal cooling ducts and speed up cooling cycles.

Dashboards, instrument panel covers, brake handles, and other plastic internal components that are injection molded are just a few examples. Complex injection mold tools require months of planning, development, and manufacturing, as well as money. Companies are therefore looking for methods to design minimally machined 3D printed injection molds. Additionally, this saves a lot of money on design and production.

Making jigs and fittings using 3D printing is very efficient and affordable. Moreover, several factory robots that handle the parts would require grippers that can perform a particular set of tasks in addition to the manual handling aspect. The handling of the subsequent production line must be changed if a new vehicle model is being created. The robotic grippers can be 3D printed, which saves money and labor. A few electrostatic discharge (ESD) materials are available for 3D printing that stop static electricity buildup on their surface, protecting electronic components from harm during assembly.

Other phase of electric car production: The aftermarket is a crucial stage of the electric vehicle manufacturing. Companies intend to alter the ecosystem by investing in 3D printing firms that can create spare components for them to guarantee seamless aftermarket services. Even though 3D printing is a slow process that cannot handle mass production volumes, businesses will profit in the future from the flexibility, lower costs, and faster delivery periods provided by 3D printing techniques. The expense of investing in 3D printer machines saves on inventory and stock keeping costs because of digital printing and intelligent inventory models.

An increase in consumer preference for electric vehicles and the availability of a variety of kinds of electric cars are both significant facts that are mirrored in the steady development of the global electric vehicle market. The increased popularity is prompting producers to create new electric car models. However, to create and improve electronic vehicles, producers have to tackle a number of obstacles. By using innovative methods, producers are able to surmount standard design and development difficulties for electric vehicles. Technology for additive manufacturing, also known as 3D printing, aids producers in overcoming some of these major difficulties.

To accelerate time to market and save money on production, top electric vehicle makers favor 3D printing solutions over conventional or subtractive manufacturing techniques. Moreover, major automotive firms spend money on R&D to incorporate 3D printing into the electric vehicle (EV) assembly procedure smoothly. Numerous companies concentrate on finding solutions to the problems posed by making electronic vehicles using 3D printing technology rather than subtractive manufacturing technology.

One of the main obstacles in the production of EVs is extending battery life. Automobile manufacturers spend money on R&D to develop better rechargeable batteries that will boost EV sales. Akin to commonly used battery technologies, 3D printing makes EV batteries last longer in a variety of ways. Modern pieces and parts can be 3D printed by engineers to lighten the weight of an EV and increase battery life. By combining various parts, they can also create lightweight electric vehicles. Engineers can also use 3D printing to create heat exchangers and encasings that control temperatures to prolong battery life.

The automotive sector will use 3D printing extensively to aid businesses gain an edge over rivals. According to the respondents, it will be a key piece of technology used to support businesses as they implement plans to boost productivity. In reality, 89% of Top Performers will consider novel design approaches to benefit from 3D printing. Top Performers are 2.1 times more likely than peers to use scan data to create parts for 3D printing, which supports 3D printing. Additionally, 60% of businesses will use scan data or will increase their use of it if it becomes simpler to use because it will increase the effectiveness of their design processes.

A 3D-printed electric car called YOYO was unveiled in 2022 by XEV, a cutting-edge electric vehicle startup business. Using 3D printing technologies, XEV is able to market its products in Italy for about €10,000 without the use of expensive machine tooling.

To avert corrosion on the aluminum surface, a primer made of green zinc chromate or zinc phosphate is used in aircraft. All newly manufactured aircraft are coated with primer and an anti-corrosion substance.3D printing has been used for production of different aviation products. Accelerated production procedures are now conceivable in the aircraft construction industry because of 3D printing. When necessary, manufacturing can be done accurately. Another benefit is that demand-driven additive manufacturing reduces waste caused by excesses, which frequently happens with serial production.

In recent years, 3D printing technology has strengthened the airplane sector. The use of 3D printing technology simplifies complex and small-scale construction. Aerospace parts such as door handles and light housings, which prioritize form above utility, are frequently produced using 3D printing. All aviation businesses are keen to invest in this field as3D printed parts can make aircraft lighter, quieter, and more effective.

Aviation is a viable sector for 3D printing to thrive owing to the strict certification requirements and demanding safety clearances. When 3D-printed parts made of super alloys and high-tech thermoplastic composites pass federal aviation regulations, there is growing confidence that additive manufacturing has a place in other extremely demanding industries.

General Electric (GE) believes that 3D printing in the aviation will lead to improved performance from fewer parts, huge cost savings per plane, significant impact on development and production of jet engines for airplane, and easy printing of spare parts for airplane.

Tooling limitations are removed via direct production from computer-aided design data. Developers and innovators have liberty to create parts with extremely intricate internal structures, and complicated rocket and jet engines are constructed from smaller parts. Sophisticated designs made feasible by 3D-printed parts can occasionally result in astonishing reductions in the number of parts needed. Given that aviation and aerospace designs frequently call for extremely complicated part assemblies, any effort to streamline assembly is appreciated. Through fewer joints requiring welding, possible failure modes are eliminated, saving time and money. To investigate the limitations of the AM technology accessible at the time, GE Aviation and Morris technology teamed up in 2012. A jet engine fuel nozzle's 20 parts were reduced to one by engineers, resulting in a weight savings of 25% and faster assembly times. Twelve 3D-printed parts replaced 855 produced by several contractors in a new sophisticated turboprop engine. The number of parts is decreased, and supply channels are streamlined as well. As additive manufacturing's maximum part sizes and build rates rise, the effect of parts consolidation will become more pronounced. Virgin Orbit, for instance, anticipates using additive manufacturing to create whole thrust chamber assemblies.

A greater embrace of 3D printing is also expected to boost productivity for manufacturers. Boeing anticipates saving between $2 and $3 million per plane by employing 3D printing to make parts for its 787s.

According to market research firm Netscribes, revenues from 3D printing processes could be $8.7 billion annually worldwide in 2020. A large portion of this increase was driven by 3D printing like direct metal laser melting (DMLM) and electron beam melting (EBM), in part because these techniques generate high-value components and usable prototypes for the aviation, aerospace, and other industries.

Jet engine prototypes and finished parts printed with 3D printers are making a huge impact on research and manufacturing. High-bypass turbofan aircraft engines from the GE90 series were first used in 1995. To operate the next model of wide-body aircraft like the Boeing 777X, the GE9X engine from GE Aviation feature 19 fuel nozzles that were 3D printed. The front fan alone on the GE9X measures 11 feet in diameter, making it the largest and most potent jet engine ever produced. Even though it uses about 10% less fuel than the previous model, it still has a takeoff thrust of 100,000 pounds.

The initial two LEAP-1A engines for Airbus's newest A320 passenger plane arrived in April 2016. The makers of the engine, CFM International, have the same partners GE Aviation and Safran of France. Nearly 12,000 requests for the LEAP engine and its 3D-printed fuel nozzles have been placed with CFM. To satisfy the growing demands, a manufacturing plant in Alabama will build thousands of nozzles needed. When using 3D printing, functional prototype testing is frequently more affordable, less complicated, and faster. The Future Affordable Turbine Engine (FATE) engine's prototype testing was finished by GE Aviation in late 2017 in partnership with the US Army. The development and production of the turbine rig for the prototype heavily relied on 3D printing. The effectiveness and performance of the FATE engine are unprecedented. It consumes 35% less fuel than existing engines in use while reducing production and maintenance costs by 45%. For Apache and Black Hawk helicopters, the ITEP engine is designed to act as a direct swap. The engine is expected to save 25% on fuel and 35% on manufacturing expenses. Functional prototypes of 3D-printed parts are used in the T901 prototype of the suggested production model. The utilization of 3D printing in the FATE and ITEP programs builds on knowledge gained from the earlier LEAP engine development for commercial usage. Select lightweight, more durable jet engine parts can be produced using 3D printing, which offers the perfect combination of streamlined design, lower costs, and faster turnaround times.

The first FAA approval for GE Aviation to deploy a 3D-printed component in a commercial aircraft engine was granted in April 2015. Utilizing a selective laser sintering (SLS) technique, the sensor housing for a compressor inlet temperature sensor is made. Over 400 GE90-94B engines for Boeing 777 aircraft will employ the component because of a contract GE signed with Boeing. Combustors for jet engines are remarkably challenging to make the old-fashioned way. Superior parts are created using a 3D printing using cobalt-chrome alloy powder as the build material to shield the temperature sensor's extremely delicate electronics from soaring airflows and considerable engine icing. Other aircraft parts, such as those used in ducting systems, are now being made using 3D printing.

For the aerospace sector, complex tools are also produced via 3D printing. A trim-and-drill tool was created in 2016 by Oak Ridge National Laboratory (ORNL) researchers to aid Boeing in producing airplane wings for its 777X aircraft. The tool measures 1.5 feet tall, 5.5 feet wide, and 17.5 feet long. The 1,650-pound device was printed out of composite polymers and carbon fiber in around 30 hours. It was later recognized as the biggest solid item made on a 3D printer by the Guinness Book of World Records.

The use of lightweight, 3D-printed items in aviation and aerospace applications results in energy savings, decreased carbon emissions, and improved return on investment (ROI).

Production for space applications, requires an exceptionally high degree of precision. EBM and DMLM, two 3D printing techniques, excel at producing items with precise tolerances. When layers are far thinner than the width of a human hair, at 20 or 40 microns, a high level of dimensional precision is possible.

8.6 CORROSION IMPACT ON 3D PRINTING APPLICATION IN SPACE

Corrosion also occurs in space. Materials deployed to space undergo exposure to vacuum, ultraviolet, and X-ray bombardment, and high-energy charged particles (mainly electrons and protons from solar wind), which replace oxygen and moisture as the main sources of corrosion. The atmospheric atoms, ions, and free radicals, most notably atomic oxygen, play a significant role in the higher layers of the atmosphere (between 90 and 800 km). Because waves of ultraviolet radiation cause molecular oxygen to photodissociate, altitude and solar activity affect the amount of atomic oxygen. The atmosphere contains approximately90% atomic oxygen around 160 and 560 km. Appropriate material selection and material technique aid in combating corrosion, with 3D printing being one of such material techniques.

There are more instances than ever of 3D printing being used in space to reach low-earth orbit and beyond. NASA researchers created and printed a working prototype of a two-piece rocket injector at the Marshall Space Flight Center (MSFC) in Huntsville, Alabama, which met the performance criteria of its 163-part predecessor. While decreasing production and assembly times, the significantly streamlined 3D-printed injector satisfies demanding mechanical property and hot fire specifications.

The application of 3D printing in space has been on for some time. This is due to the unique advantage it offers for the space industry. Based on the special geometries that 3D printing enables along with the materials utilized, it allows the production of lighter, less expensive, and more effective pieces. A lot of testing and research is being done to accelerate deployment of 3D printing in space. To determine if Incus' lithography-based metal manufacturing technique would be practical for producing parts in a lunar base utilizing scrap metal or already-existing surface materials, Incus and the European Space Agency (ESA) have teamed up together. Additionally, experiments are being conducted on the ISS to see whether bioprinting will be useful in the future in space.

It is being used as a replacement for building structures including shelter in space, as well as to produce space rockets, spare parts, and food that can be consumed in space. In space, 3D printing is facilitating the production of replacement tools and components in orbit. However, smooth 3D-printed rocket engines have been in great supply since the 2000s. For instance, the Rutherford rocket engine from Rocket Lab employs injectors, pumps, and combustion chambers that were manufactured using an EBM technique.

3D printing has the potential to significantly reduce the logistics requirements for the International Space Station and upcoming long-duration human space missions. More significantly, there are components that astronauts can print in zero gravity or space. Furthermore, 3D printing has a number of benefits that advance space exploration.

Whether creating satellites or any other type of equipment in space, 3D printing still constitutes a cutting-edge technology. Some believe that 3D printing will disrupt industries, but this is untrue. Since quite some time 3D printing has been deployed not just in the aerospace sector.

3D printing has been employed in an array of projects by businesses such as Boeing and Airbus to produce increasingly complicated, lighter parts for their satellites. For instance, last year, MIT 3D printed ion-powered nanosatellite thrusters, while Airbus used 3D printing to produce radio frequency (RF) parts for two Eurostar Neo satellites. Australian company Fleet Space revealed that it was planning to send 3D-printed satellites to orbit alongside the Centauri constellation. ICON and Big-Bjarke Ingels Groups for NASA have finished the most challenging project in the area; they produced a lifelike simulation of a 3D printed dwelling, known as Mars Dune Alpha. The objective is to "transport" it to Mars to enable extended space exploration expeditions.

3D printing is also being used to produce thrusters for space. It was used to produce Griffin Lunar lander called Attitude Control Thrusters (ACTs) by NASA in partnership with Agile Space Industries. The ACTs was designed to thrust Volantines Investigating Polar Exploration Rover (VIPER) to the south pole of the moon in 2023. It will also be used to map and detect points in the moon where ice is found.

There was announcement of a proposal by the Israeli company Aleph Farms to 3D-print beef inside the International Space Station (ISS) in 2019. They were able to produce cell-based meat that has the same taste, texture, and composition as a traditional steak using a Bioprinting Solutions' 3D printer. Bovine cells were taken from cows on earth and flown to space, where they were cultivated, to accomplish this. Aleph Farms has also taken part in the so-called Mission: Space Food. An integrated plan for human nutrition in space is being developed by a group of professionals in the fields of technology, food, and space. The success of this will create novel space cuisine using the Astrea company's engineering and cooking prowess to enable astronauts flourish in this environment. NASA is also working with other companies such as BigRedBites, ALSEC, Electric Cow and KEETA to 3D-print food in space.

3D printers have been used to replicate SpaceX's 3D-printed space suits and helmets. Each helmet is equipped with visors, locks, valves, and microphones, and the suits are appropriate for space flight. This clothing was produced using the FDM method of printing the helmets as it offers a greater selection of cutting-edge materials, like PEKK.

8.7 CORROSION IMPACT ON CONSTRUCTION IN 3D PRINTING APPLICATION

The size of the global construction market reached about USD 12.74 trillion in 2022 [11]. This represents about 6% of the global GDP. Indeed, the world's economy and prosperity depend heavily on engineering and building. Therefore, construction companies are constantly searching for innovative ways to boost capacity while cutting down expenses. One of the most recent innovations in the field of construction technology is 3D construction. The construction industry is being propelled toward automation to reduce expenses, boost output, enhance architectural freedom, and lower labor expenses. Additionally, 3D printing helps save the environment.

In the construction sector, 3D printing was first used experimentally in the late 1990s. Modern procedures (such as contour crafting, D-shape, and concrete printing), materials (such as cementitious, polymer, and metal), and deployment methods (such as off-site/on-site fabrication, hybrid approaches, and multiple materials) are the main topics of current 3DP research. Some of the construction applications include in situ repair, bespoke parts, novel shapes, and topology optimization.

There are various forms of 3D printing used in building. Four principal 3D printing techniques among them are contour crafting, D-shape, concrete printing, and shotcrete, as depicted in Figure 8.9.

i. *Concrete printing*: This type of 3D printing for construction uses digital manufacturing procedures for cementitious materials, also referred to as "3D concrete printing" or "concrete printing." It has a plethora of applications. There are numerous other business prospects, including block walls, sheds, garages, and monument lettering, but the majority are only interested in implementing the cost and time reductions in residential home development. Large machinery is used in this printing method, which extrudes cement-based materials from a huge nozzle. To create the required architectural element, these 3D printers are frequently designed as gantries

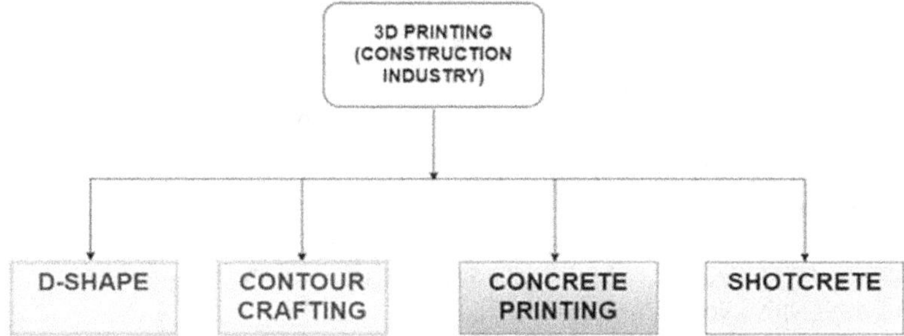

FIGURE 8.9 Classifications of 3D printing for construction.

or robotic arms that continuously extrude a dough-like material layer by layer. Three steps – data preparation, concrete material preparation, and 3D printing – make up the conventional concrete 3D printing process. A 3D printer for building might cost anywhere from $180,000 to over $1,000,000. Price-wise, robotic arm systems are often more expensive than gantry-style systems.

ii. *Contour crafting*: The most promising 3D printing method used in the construction sector is called contour crafting (CC). With this method, the material is introduced one layer at a time, yet the whole thing happens on-site. By deploying a 3D printer capable of manufacturing an entire house on-site, this innovation presents a significant opportunity for the automation of the construction procedure.

iii. *Shotcrete:* Shotcrete was invented to mitigate the issues (weak interlayer bond and challenges in building) which extrusion-based 3D concrete printing method experiences. Shotcrete, also known as gunite or sprayed concrete, is a building technology devised by Carl Akeley in 1907 in which concrete or mortar is delivered by a hose and pneumatically shot at high velocity onto an area. Usually, it is strengthened with regular steel rods, steel mesh, or fibers.

The Digital Building Fabrication Laboratory (DBFL) of Braunschweig University of Technology is where shotcrete concrete 3D printing was initially made. Shotcrete can be used in two fundamental ways: dry-mix process and wet-mix process, with the key distinction being the timing of the addition of water to the mixture. Wet-mix shotcrete tends to generate smaller amount of dust and rebound than the other two, making it the procedure of choice for engineers. Shotcrete is typically controlled by nozzle men, although recently, the use of robotic arms in construction has made it possible to computerize the shotcrete process.

The location and material characteristics are also significantly affected by the rebound of particles or fibers [12]). Even more challenging than the creation of extrusion-based 3D printable mixtures, the shotcrete technique also necessitates thorough evaluation of the printing mixtures and their practicality to guarantee suitable pumpability and shootability at the same time. Compared to extrusion-based 3D printing, shotcrete concrete 3D printing also requires intricate robot control to obtain the appropriate printing geometry. The ultimate geometry and quality of the printed structures are considerably influenced by a variety of factors, such as the pipe configuration, pumping pressure, time between following layers, and nozzle size, shape, and travel speed. In general, achieving high geometry precision for SC3DP is more challenging than it is for extrusion-based 3D printing.

iv. *D-shape:* This type uses a non-hydraulic Sorel cement as binder, sometimes called binder jetting 3D concrete printing. Magnesium oxide activates the sand in the powder bed. Binder used is aqueous magnesium chloride. It is named after the demonstrator called Enrico Dini with D-shape.

It is used for large architectural piece, for furniture, and bridge, among others. D-shape was used by KOL/MAC LLC architecture for producing coffee table, root chair in 2009. Shiro studio used it for Radiolaria pavilion in 2008. Acciona designed a footbridge with it in 2017.

D-shape advantage lies in the ability to create hollow parts, high flexibility, and creation of unsupported overhangs. It does not require auxiliary support as unbounded powder gives the support to next layer during printing. Key limitation of the D-shape is lack of reuse of the cement and aggregate powder because of the presence of ambient humidity causing hydration. Hence, it is used in off-site.

Concrete 3D printers technology: Cable drive system, robotic arm, gantry robots are the key technologies used in concrete 3D printers. The technology used is influenced by magnitude of the print, technique to be employed, and application. The modelling software such as Revit or proprietary 3D printer software feeds the printer with structure plans. Thereafter, concrete 3D printer head fitted with nozzle extrudes the concrete ably supported by a structure or platform.

i. *Crane or cable-driven technology:* Crane or cables are used to support concrete 3D printer head. It is easy to transport, being lightweight with several geometric freedoms than gantry. Key limitation is the need for space for the crane or cabling to prevent overlapping with the printer.

ii. *Robotic arm technology:* This has optimum 3D printing freedom and six-axis movement. It is used for pouring concrete, component embedding (rebar), post-processing operation. It is a compact technology used for small-scale operations. They require crane-operator expertise to be able to control and operate it.

iii. *Gantry technology:* This technology has moveable gantry system that can mix and pour concrete. It uses vertical extrusion with good scalability and best stability. Limitation is the associated cost (set-up and transportation) as it needs to be larger than the assemble structure. There are different sizes (small to large) requiring little expertise for usage and control. Figure 8.10 summarizes major producers of concrete 3D printers around the world.

COBOD produced in Denmark is the fastest and obvious choice for many users that employ gantry technology. Italy is home to Vertico that uses robotic arm as the printer type and another company known as WASP that uses cable-driven and gantry technology. Construction-3D is based in France and uses crane and robotic arm. ICON based in the USA uses gantry technology.

Pan et al. [13] summarized the technological details of four standard concrete 3D printing techniques as shown in Table 8.2.

Materials requirement for 3D printing for construction: Materials for 3D printing must be comparatively low in yield stress and plastic viscosity to enable optimal flowability during pumping. As for "buildability," that is, the ability to stack layers on

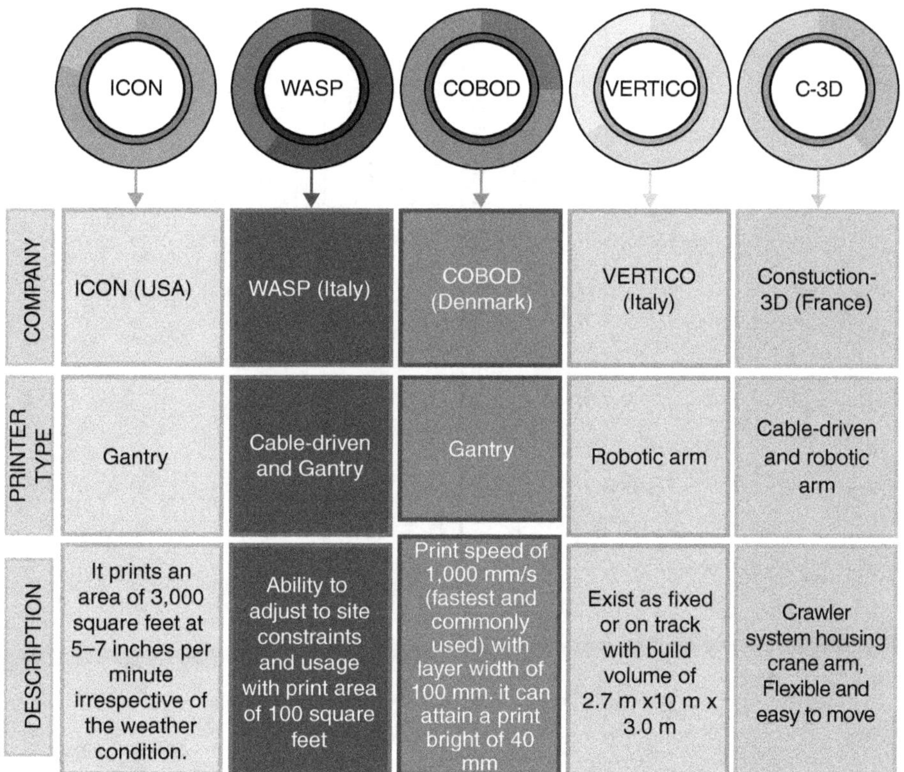

FIGURE 8.10 Summary of concrete 3D printer producers.

top of one another, the materials need to be less flowable and demonstrate a greater plastic viscosity [14]. Another trade-off is the association between printability and open time (interval between printing a document and setting it) [15]. Most popular building 3D printing material is Portland cement concrete due to its unique qualities that are well-known.

Various materials have been employed as raw material for 3D printing for construction industry as depicted in Figure 8.11.

They are grouped into organic, non-organic and composites [13]. Non-organic components included metal, sand, recovered construction debris, cementitious materials, space materials, and sea salt. The organic materials that were utilized were polymer, wood, and algae. Composite materials included soil, polymer/metal, rock and thread, and soil. More than half of the investigations employed cementitious materials, followed by polymeric and space materials. Wood, soil, recovered building waste, and sand were all utilized at least three times.

Concrete 3D printer parameters: Quality of a print is affected by the print head design, printer type, print speed, pumps and controls, concrete mix design. A longer print time reduces interlayer bonding, a decrease in nozzle speed increases the adhesive strength. Printing process obstructions could be caused by short distance

TABLE 8.2
Four 3D construction printing technological details

S/N	3D printing type	Topology	Material	Layer depth	Location	Support	Head mounting
1.	D-shape	3D	Powder deposition	4–6 mm	Off-site	Yes	Gantry based
2.	Contour crafting	2.5D	Extrusion (paste)	~13 mm	On-site	No	Cable-driven
3.	Concrete printing	3D	Extrusion (paste)	4–6 mm	Off-site	Yes	Gantry based
4.	Shotcrete	3D	Extrusion (paste)	20 mm	Off-site	No	Robotic arm

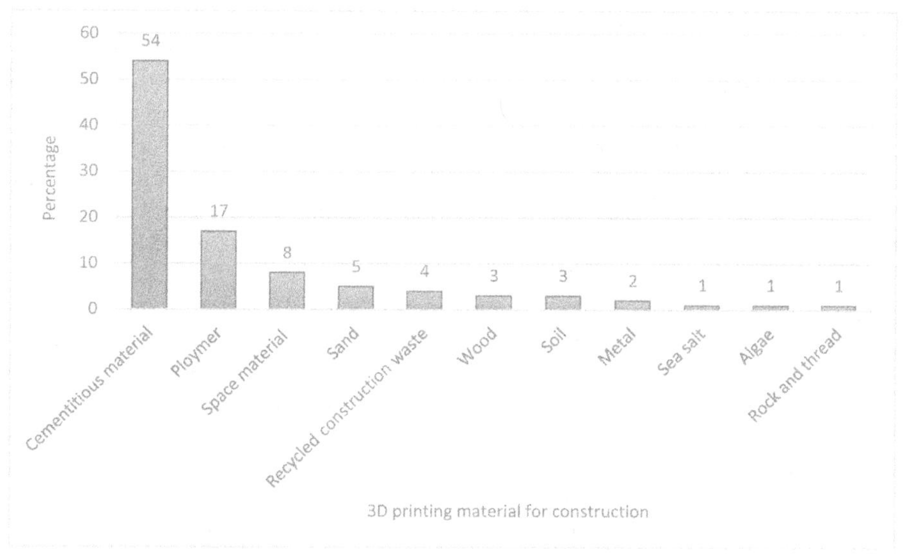

FIGURE 8.11 A graph of 3D printing materials used for construction.

of nozzle to the print surface. Best performance is obtained with a concrete print height being equal to nozzle width. A circular nozzle is best suited for complex geometries. However, a rectangular nozzle gives higher strength due to fewer gaps in printed layers. Extrudability and buildability are the key mix design properties. A balance between workability and stiffness needs to be achieved for extrudability and buildability. Each layer needs to bond properly. A stiff mix may result in print head clogging. Optimum workability and extrudability are achieved by reducing the stiffness but with reduced strength and poor buildability. Supplementary cementitious materials (SCMs) used for concrete 3D printing include fly ash, metakaolin, silica fume, and superplasticizers.

Challenges of concrete 3D printing: The 3D printing material presents the most difficulties for the construction industry. Printability, buildability, and open time are the three categories into which the material challenges can be categorized. To be extruded from the nozzle and maintain its shape, the material must have the desired buildability and printability.

Economic and environmental impact of concrete 3D printing: It boosts productivity and the speed of completion (indirect economic benefit) as geometric complexity increases. There is no need for any formwork. Conventional concrete construction is made up of 50% form work. In terms of environmental impact, concrete 3D-printed structure depends on the process, material, and geometric complexity. Formwork removal from concrete 3D printing produces less materials needed, less wastage, and less environmental impact. A lifecycle assessment of conventional and concrete 3D printing shows that concrete 3D printing poses less danger to the environment when there is no reinforcement used. Additionally, concrete 3D printer poses less risk to environment as geometry structure complexity increases.

Projects implemented with concrete 3D printers: Approximately40% of homeless population in Texas, USA, were accommodated in 2020 by a project anchored by ICON. A 400-square foot home was built in 27 hours using Vulcan 3D printer by ICON. Another project that will cost clients 400,000 US dollars in Georgetown, USA, is being planned by ICON for last part of 2023. Vulcan 3D printer by ICON can produce eight floor plans of three or four bedroom with two or three bathrooms. This is possible because the ICON-developed Lavacrete concrete mix, which can adapt to site the environment and instantly deliver read-to-print concrete, is fed to the Vulcan printer via the Magma concrete feeding system.

By utilizing the COBOD BOD2 printer and the concrete mixture from Heidelberg Cement, PERI built the initial 3D-printed home in Germany in September 2020. A plant printed 24 concrete components, which were then shipped to the assembly location. In November 2020, the 160m^2 home was finished by COBOD printer, which produced 1 m^2 of wall each 5 minutes. Printing the walls, which included placing the water, power, and pipe connections, only took two workers.

The world's biggest non-profit organization dedicated to building houses, Habitat for Humanity, planned to construct two 3D-printed houses in Williamsburg, Virginia, and Tempe, Arizona, by 2021. The 1,200 square foot Virginia home was created using a COBOD 3D printer in about 28 hours, which is about four weeks quicker than conventional building. According to the corporation, the 3D-printed concrete walls reduced building expenses by 15%per square foot. The 1,738 square foot home in Arizona was built over the summer, when work is generally put on hold because of the intense heat. The interior and exterior walls, as well as 80% of the total structure, were built utilizing 3D printing. The remaining 20%, such as the roof, were built using conventional techniques.

The longest 3D-printed concrete pedestrian bridge in the globe, measuring 29 meters, was unveiled in the Dutch city of Nijmegen in 2021. Because concrete was only used where it was necessary for structural strength, it was projected that 3D printing resulted in material savings of about 50%. BAM and Weber Beamix produced

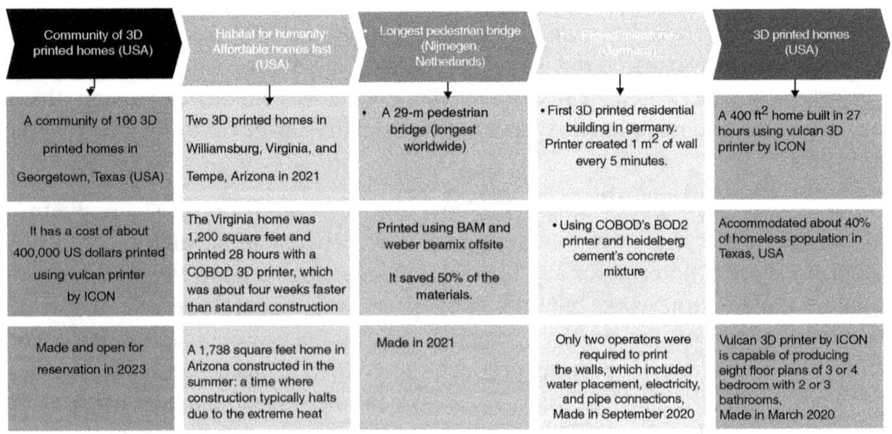

FIGURE 8.12 A summary of concrete 3D printed key projects.

the 3D-printed bridge sections off-site, which were subsequently transported and put together on-site. The longest 3D-printed concrete bridge to date was 26 meters long and was built by Tsinghua University in Shanghai. These projects are summarized in Figure 8.12.

8.8 CORROSION IMPACT ON WARFARE IN 3D PRINTING APPLICATION

The military can become more autonomous using 3D printing. Armed forces no longer need to rely on third-party vendors for equipment replacement parts and components because of 3D printers. In a time of conflict when enemy forces would try to cut off supply lines, this might be a crucial skill. The military is constantly seeking for novel ways to remain abreast of the curve, and one such approach is through integrating 3D printing into the defense industry. Globally, military troops from many nations use 3D printing. In addition to innovative designs for the safety gear used by service troops, spare components for essential vehicles, ships, and airplanes, prototypes for novel defensive weaponry, medical, bunker.

The fact that 3D printing enables on-demand production, one of its main benefits is that militaries can create the products they require whenever they require. In combat scenarios where speed is of the essence and conventional supply channels are disrupted, this could prove particularly helpful. Upcoming military use of 3D printing will increase, as the technology is utilized to create things from medical and food items to weapons and ammo.

3D printing has been deployed in spare parts production for military, creating tools and jigs, to improve and maintain military equipment such as antenna housing, production of medical devices and military supplies, 3D printing of military barracks, bridges, 3D printing of propellers, vehicle hulls, prototyping safety wears and equipment.

Military uses 3D printing to manufacture spare components for ships, airplanes, and ground vehicles. When a thing is required as a substitute component in the past, it had to be procured from a manufacturer, sent to the military base, or flown in for a ship that was at sea. Given the present logistical challenges affecting the globe as a whole and keeping crucial parts stranded in ports as a result of chip shortages and backlogs brought on by the pandemic, this might take weeks or even months. The military can use 3D printing to produce the parts they require on-site, whether that is at their base, the front lines, or at sea. This saves cost and time that would hitherto be spent in ordering a custom part. It is also used in creating jigs and fixtures for repair or install equipment utilized by the military to develop novel weapons and make prototypes. The military gains from this technology as it makes it possible to produce prototypes and modify designs more quickly. Furthermore, 3D printing is utilized to make special components that meet the equipment's requirements and enhance workflow.

Creating housings for pre-existing equipment aims to shield them from damage during usage. An antenna was built with a robust 3D-printed housing to prevent snapping off near the base during intensive use and traveling in dense brush. It is used to enhance military personnel's workflow, safeguard current equipment, and guarantee an extended lifespan for those items of equipment as they are used on a daily basis. Medical equipment and supplies are also made with it. In battle scenarios, where time is crucial, this is particularly essential. Soldiers can receive treatment more rapidly and have a better chance of surviving combat circumstances because of the quick and simple production of 3D-printed medical items on the field. World over, military-related organizations are coming up with endless methods to produce healthcare equipment using 3D printing technology, from specialized Epi-Pen carrying cases and stethoscopes to intravenous (IV)line control, tracheal tubes, splints, casts, covid swabs, face shields, earplugs, and so much more.

Military is constructing 3D-printed barracks to shelter service troops on training missions in collaboration with businesses like ICON. Additionally, military uses 3D printing to construct buildings, bridges, vehicle hideouts, and bunkers all over the world to support and house military equipment and soldiers both at bases and in the field. Huge robot 3D printers that extrude concrete are used to build these massive structures. Following a natural disaster or a mission during a conflict, buildings, bridges, and other structures might be left behind for local residents to use once the military use has ceased. Moreover, 3D-printed runway panels are being developed for use in regions where military aircraft must land to carry out their missions or deliver supplies to ground forces but where the runways are unstable and too dangerous for personnel to land safely.

Five 200-kg blades make up the propeller that the French business Naval Group 3D printed. The minesweeper Andromeda eventually received the propeller. They have significantly cut the number of materials needed for this component's manufacturing and building time by employing 3D printing technology, saving thousands of dollars in the process. The US Army and the Applied Science & Technology Research Organization (commonly known as ASTRO America) have been collaborating on the Jointless Hull Project to create and distribute a combat vehicle hull. They anticipate that this effort will aid in speeding up production and cutting costs. The vehicle's efficiency and survivability for the military men operating it will also improve due to the reduction in weight.

The U.S. Army and General Lattice have teamed up to use 3D printing and innovative lattice geometries to enhance the damage absorption of the Army's combat helmet, with the goal of increasing the field protection of soldiers and their chances of surviving a head blow. To refine the design and eventually start putting it to use in the field, the 3D-printed materials will be put to the test in actual circumstances. The capacity to 3D-print a working model and make changes quickly is essential more than ever when redesigning designs for face shields, masks, and other items of safety equipment.

8.9 CORROSION IMPACT ON FOOD IN 3D PRINTING APPLICATION

A growing population means increase in demand for food. Vegetarians and non-vegetarians all need food to survive. Numerous efforts have been made to produce food using different techniques. 3D printing has been adapted for food production. Chocolate and other food ingredients have been adapted for 3D printing of food. NASA 3D-printed a pizza. Research is ongoing to extract food ingredients from seaweed, lupine, agricultural and food waste to be used as feedstock for 3D printing. Additionally, the emission occasioned by methane from meat is being replaced through 3D printing of meat without depleting natural animals and arable land.

Akin to ordinary printing, printing food requires the printer to be modified to work with nutrients such as sugar, starches, and proteins rather than ink. To manufacture unique processed foods, 3D bioprinters are used to layer minute, semi-liquefied food fragments precisely on atop one another.

This method of obtaining refined food offers several benefits, including meal composition that is tailored to the individual diet, simplicity of transportation to even the most remote locations on Earth or into space, and a longer shelf life [16]. NASA started looking on 3D-printed meals in 2006. The Chef3D was created in 2013 by the NASA Advanced Food Program with the goal of promoting a healthier diet for the astronauts' team on longer trips. A pizza might be printed in 3D using this bioprinter and then simply baked.

Solid Free-Form (SFF) is the foundation of 3D printing for food. In addition to alternative techniques such as stereolithography laser sintering (SLS) and selective laser sintering (SLS), the main 3D printing technique that can be used to print food is called FDM. The FDM method relies on directly layering purees, gels, and doughs without making use of a structural agent to help maintain their structures. Various containers hold the combined materials. The ingredients can be delivered by pressing them onto a surface with an electric motor attached to a syringe that is coupled to one or two nozzles. A device known as a dual-feed extruder forces two materials of distinct colors out of the nozzle to produce a third color by varying the mixing ratio managed by a colormix generator [17].

Revo Foods, an Austrian startup, wants to 3D-print many seafood products, including salmon slices, using different plants. Even though Revo Foods presently sells products in the shape of spreads that have not yet been 3D printed, the business aims to use 3D technologies in the near future to build a robotic manufacturing facility and 3D-print more salmon in all its varieties. Nutritionists in Chile have created dishes made from algae using 3D printing. These dishes were created from Cochayuyo algae, a plant indigenous to Chile that is a significant source of plant protein, and are loaded with vital nutrients and appealing designs.

8.10 SUMMARY

Corrosion is an ancient and modern technology that has attracted a lot of research and investment. It occurs in almost every sector of the economy with devastating impact if not checked. Several techniques have been employed to curb and reduce the impact. This chapter examined the corrosion in relation to an emerging technology known as additive manufacturing (with emphasis on 3D printing). Although most materials corrode, selecting the appropriate material reduces or prevents corrosion. 3D printing is not exempted from corrosion as conventional materials are used for 3D printing operation. However, the history of 3D printing spans several decades starting from 1950 to present, but it gained prominence a couple of years back. A brief discussion on 3D printing, filaments, and method was examined, where it was established that FDM and sintering are the most used and the latter the oldest.3D printers are also being developed to print a variety of materials, including plastics, metals, composites, and many more. When it comes to industrial 3D printing, there are many materials to choose from.3D printing enables on-demand production, as well as being flexible, customizable, and having other great advantages that make it a technology of choice. The application of 3D printing in the major sectors where corrosion is prevalent was examined in this chapter. Applications such as food, aviation, transportation, and construction are affected by corrosion. 3D printing has evolved in those these sectors with a view of improving them and also reducing corrosion in them.

REFERENCES

1. Park, S.J., Lee, J.E., Lee, H.B., Park, J., Lee, N.-K., Son, Y. and Park, S.-H., 2020. 3D printing of bio-based polycarbonate and its potential applications in ecofriendly indoor manufacturing. *Additive Manufacturing*, *31*, 100974.
2. Pereira, T., Barroso, S. and Gil, M.M., 2021. Food texture design by 3D printing: A review. *Foods*, 10(2), p.320.
3. Esmond, R.W. and Phero, G.C., 2015.The additive manufacturing revolution and the corresponding legal landscape: This paper discusses the ways to protect innovations in additive manufacturing in this fast changing world. *Virtual and Physical Prototyping*, *10*(1), pp.9–12.
4. Brinson, H.F. and Brinson, L.C., 2015. Characteristics, applications and properties of polymers. In *Polymer Engineering Science and Viscoelasticity* (pp. 57–100). Springer.
5. James, M.B., 2019. Polymers in civil engineering: Review of alternative materials for superior performance. *Journal of Applied Science and Computations*, *6*(5).
6. Egan, P.F., Bauer, I., Shea, K. and Ferguson, S.J., 2019. Mechanics of three-dimensional printed lattices for biomedical devices. *Journal of Mechanical Design*, *141*(3).
7. Chantarapanich, N., Puttawibul, P., Sitthiseripratip, K., Sucharitpwatskul, S. and Chantaweroad, S., 2013. Study of the mechanical properties of photo-cured epoxy resin fabricated by stereolithography process. *Songklanakarin Journal of Science and Technology*, *35*(1), pp.91–98.
8. Cherdo, L., 2022. *Metal 3D Printers in 2022: A Comprehensive Guide*. AIE Publisher.
9. Zhang, J., Wang, J., Dong, S., Yu, X. and Han, B., 2019. A review of the current progress and application of 3D printed concrete. *Composites Part A: Applied Science and Manufacturing*, *125*, 105533.
10. Eliaz, N., 2019. Corrosion of metallic biomaterials: A review. *Materials*, *12*(3), p.407.
11. Awuah, K.G.B. and Abdulai, R.T., 2022. Urban land and development management in a challenged developing world: An overview of new reflections. *Land*, *11*(1), p.129.
12. Heidarnezhad, F. and Zhang, Q., 2022. Shotcrete based 3D concrete printing: State of art, challenges, and opportunities. *Construction and Building Materials*, *323*, 126545.
13. Pan, Y., Zhang, Y., Zhang, D. and Song, Y., 2021. 3D printing in construction: State of the art and applications. *The International Journal of Advanced Manufacturing Technology*, *115*(5–6), pp.1329–1348.
14. Nerella, V.N., Hempel, S. and Mechtcherine, V., 2017. Micro- and macroscopic investigations on the interface between layers of 3D printed cementitious elements.

Paper presented at the International Conference on Advances in Construction Materials and Systems, Germany.

15. Le, T.T., Austin, S.A., Lim, S., Buswell, R.A., Gibb, A.G.F. and Thorpe, T. 2012. Mix design and fresh properties for high-performance printing concrete. *Materials and Structure*, *45*(8), pp.1221–1232.
16. Izdebska, J. and Zolek-Tryznowska, Z., 2016. 3D food printing –facts and future. *Agro FOOD Industry Hi Tech*, *27*(2), pp.33–37.
17. Yang, F., Zhang, M. and Bhandari, B., 2017. Recent development in 3D food printing. *Critical Reviews in Food Science and Nutrition*, *57*, pp.3145–3153.

9 Corrosion, Industrialization, Climate Change and Energy Security in the Global South

9.1 INTRODUCTION

Unstable, expensive nature of electricity supply is affecting the Global South adversely. Companies such as PZ, Toyota, and Peugeot Assembly plant have been forced to relocate to other countries owing to high cost and unavailability of energy to break even. Energy security is needed for industrialization to succeed. However, industrialization is the key to wealth and better living, and it affects environment and ultimately contributes to climate change. Therefore, there is need to examine the interplay of energy security and climate change and the impact on industrialization. This theme forms the essence of this study.

Global South's energy is dominated by fossil-based energy. Premium motor spirit (PMS), diesel, are kerosene are major sources of greenhouse gas. The transport as well as domestic and industrial sectors of Global South rely heavily on PMS, diesel, kerosene, and wood for their energy needs. The effect of these sources on climate has resulted in massive flooding, unpredictable weather pattern, and fire incidences among others. Worse is the high cost, unstable nature, and unavailability of these sources of energy in Global South, resulting in several businesses either collapsing or moving to neighboring countries. Domestically, this fossil-fuel has caused death, respiratory diseases and untold hardship to Global South populace.

However, Global South geographic positioning favors deployment of clean energy sources. Energy sources such as solar, wind, and hydro turbine are categorized as clean energy, as they are not harmful to the ozone and the environment. They exist in abundance in Global South. Global South enjoys over seven hours of sunlight in over seven months of the year across the country. A noonday sunshine is enough to power the earth. This means that solar energy alone, if fully deployed, can solve the energy needs of Global South. Similarly, a hydro turbine requires water body (natural or artificial). Global South is home to several natural water bodies such as rivers and dams capable of providing enough energy for Global South. Other clean and renewable energy sources such as nuclear, tidal, and biomass can also be deployed

DOI: 10.1201/9781003441151-9

in Global South owing to their abundance These sources, if fully utilized, provide energy security, reduce greenhouse gases, and result in massive growth of industries.

Despite the huge potential of Global South to deploy clean energy sources, other factors continue to hinder the countries' quest for energy security and industrialization. The appropriate tools for research and development on clean energy and climate change are missing in Global South. There is limited fund for research in Global South. Countries like the USA mandate industries to liaise and fund the academia and research institutes to solve key challenges inherent in it. South Africa established a funding board that pools resources (fund and manpower) to solve its key issues. Global South currently does not have any of such resource pools. The successful deployment of clean energy requires ownership through improvement. Additionally, the enabling policy that supports and drives industrialization is not sufficient. These policies will, among other things, ensure stable, affordable, and easily available supply of energy for industrial and domestic use. It will ensure the energy becomes a contributor to climate change with emphasis on clean and green energy.

This study will examine the key concepts, trends, framework, issues, opportunities of industrialization, climate change, and energy security in Global South.

9.2 BRIEF ON THE GLOBAL SOUTH

The Global South is a region or group of countries with common economic characteristics. It is used to refer to a collection of nations based on their socio-economic and governmental traits. These regions are not grouped in the same geographical region but denote countries with status of developing or emerging countries. The term "Global South" now encompasses not only regions formerly known as the "Third World," but also regions in the North that are marked by exploitation, oppression, and neocolonial relations, such as indigenous and black communities (and immigrant communities) in Western societies, and vice versa, including Oceania, Asia, Africa, and Latin America. Many nations in the Global South are marked by low income, a large population, inadequate infrastructure, and frequently being marginalized politically or culturally.

The Global South's nations have been characterized as just becoming industrialized or in the midst of doing so, and many of them have a past of colonialism by Northern, typically European, states. The deterioration of political sovereignty has been the primary political effect of globalization on the Global South, particularly when it comes to enacting macroeconomic policies that are consistent with the needs of the Global South nations for advancement. The Southern nations with the greatest populations and economies are Brazil, China, India, Indonesia, and Mexico.

A discussion on climate change, industrialization, and energy security in the Global South is examined. Moreover, the impact of corrosion on industrialization and energy security in Global South are discussed below.

9.2.1 Climate Change

Climate change is changing the way humans live and interact with the environment. There has been increased flooding, hotter temperature, and more unpredictable weather patterns. All these factors are influencing every sphere of the ecosystem. In a clear term, climate change alters the temperature and the weather process of the earth. Despite these visible impacts of climate change, there have been divergent views about the existence of climate change. Furthermore, the level of commitment to reducing the impact of climate change differs across nations, developing and developed countries, and across sectors.

Long-term changes in temperature and weather trends are referred to as climate change. These changes may be organic, but since the 1800s, human activity has been the primary cause of climate change. This is mainly because burning fossil fuels, such as coal, oil, and gas, creates gases that retain heat.

The causes of climate change vary. The heat from the sun is trapped on Earth because of greenhouse gas pollution. Global heat and climate change result from this phenomenon. The rate of global warming is now higher than it has ever been. Burning fossil fuels such as coal, oil, and natural gas to produce electricity and heat accounts for a sizable portion of world emissions. Only about a quarter of the world's energy is still made from renewable sources such as wind and solar. The majority still comes from fossil fuels. Emissions from manufacturing and industry are primarily the result of burning fossil fuels to create energy for the production of products such as clothes, electronics, plastics, cement, iron, and steel. Gases are also released during mining and other industrial operations. Clearing forests to make way for farms, pastures, or for different uses increases emissions because when trees are cut down, the stored carbon is released. Destruction of trees reduces nature's capacity to maintain emissions from reaching the atmosphere because they absorb carbon dioxide.

The effect of climate change occurs as shift in weather patterns. Weather patterns are shifting because of warming temperatures, which is also upsetting the natural order. This puts both people and all other kinds of life on Earth in grave danger. More hot days and heat storms are occurring almost everywhere on land, and 2020 was one of the hottest years ever recorded. Rising temperatures can make it harder to work and move around, as well as raise the risk of diseases associated with heat. When the weather is hotter, wildfires develop more readily and spread more quickly. Rainfall varies in response to temperature changes. Storms become more severe and frequent consequently. They consume billions of pounds and damage homes and towns by causing flooding and landslides. In more places, water is becoming more limited. Devastating sand and dust storms that carry billions of tons of sand across countries can be triggered by droughts. As deserts grow, there is less space for agriculture. The danger of regularly not having sufficient water affects a lot of people today.

Climate change will continue to cause devastating effect if global consented effort and solution is not implemented forthwith. Earth is said to have warmed up by 1.1°C caused by greenhouse gas emission. About 23 developed countries account for 50% of all historical global CO_2 emissions, whereas about 150 countries are responsible for the other 50%. Least developed countries account for about 1.1% of global CO_2 emission. However, developing countries have lower emissions but are the worst hit by climate change. Developing countries suffer from floods and rising temperatures (heat waves) among others. Developing countries are susceptible to affect of climate change because their economy relies mainly on climate-reliant sectors such

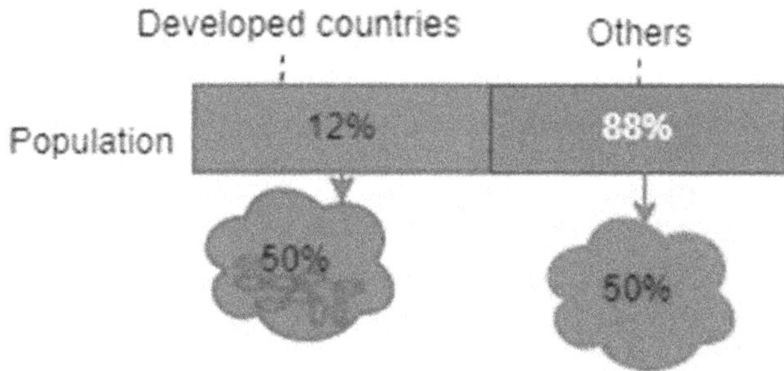

FIGURE 9.1 Global population and big CO_2 contributor.

as farming, oil and gas, and mining. There is the need for developing countries to develop strategies, solution, and implement policies and take climate actions.

A sovereign state that is considered to have a functioning economy, a high standard of living, and a sophisticated technological infrastructure is referred to as a developed country. However, underdeveloped countries are those that have the lowest socio-economic development indices and the lowest human development index (HDI) ratings. However, developing countries have a less developed industrial base and a relatively low HDI compared to the established ones. Developed nations such as USA (24.6%), Germany (5.5%), United Kingdom (4.4%), Japan (3.9%), France (2.3%), Canada (2%), and countries in Western Europe contributed to 50% CO_2 emission for the past 170 years yet having 12% of the world population as depicted in Figure 9.1.

The Brazil, Russia, India, China and South Africa (BRICS) nations which represent an emerging economy alongside other 145 nations contribute another 50% to global CO_2 emissions. China leads the pack with about 13.9%, followed by Russia (6.8%), India (3.2%), South Africa (1.3%), Brazil (1%), and Ukraine (1.8%). Least developed nations (about 47 nations) contribute 1.1% of the global CO_2 emissions.

In 2010, the superpower countries promised to support other economies with about 100 billion USD per year to enable them combat climate change before 2020. More than a decade later, they still lag with about tens of billions USD per annum. They still struggle to meet the aid need of those developing countries to mitigate the impact of climate change. This made the poorer countries to ask for a fund known as loss and damage fund from the superpowers that contributes a huge amount of the emissions. Only Scotland agreed and pledged about 2.7 million USD in November 2021 [1]. Figure 9.2 presents the chart of global carbon dioxide emission.

China leads the CO_2 emission with over 10 Gt CO_2 emitted. Countries in the Global South such as South Africa and Brazil have the least CO_2 emitted in comparison with the Global North.

9.2.1.1 Impact of Climate Change on Some Key Sectors

Figure 9.3 shows a plot of data obtained from international energy agency (iea.org) for worldwide carbon dioxide emissions of various sectors.

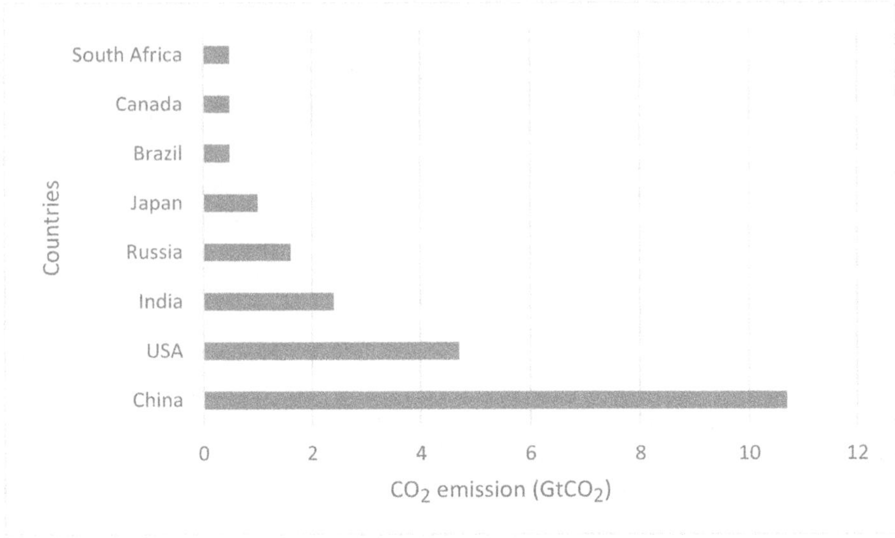

FIGURE 9.2 A chart of global CO_2 emissions of developed and developing countries.

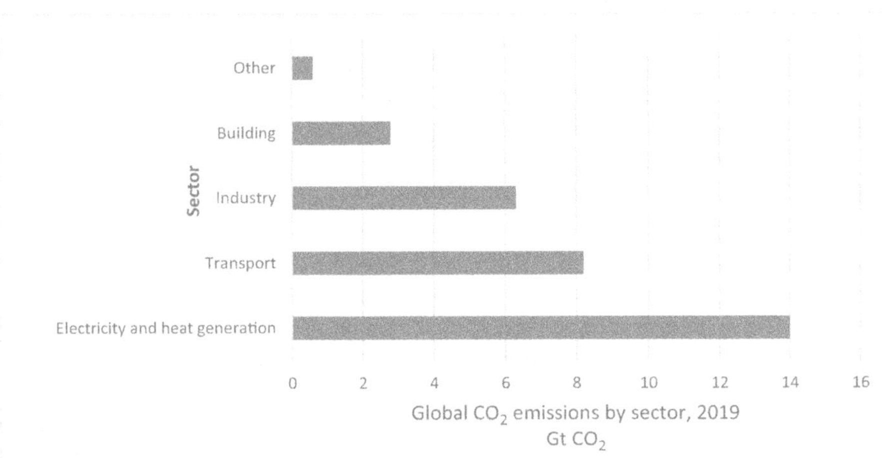

FIGURE 9.3 Worldwide CO_2 emissions by sector.

Electricity and heat generation have the highest share with about 14.0 Gt CO_2. Building has the lowest CO_2 of 2.8 Gt CO_2.

The impact of climate change is discussed here under some key issues.

9.2.1.1.1 Impact of Climate Change on Energy and Transport

Fossil fuels, which include coal, oil, and gas, undoubtedly constitute the primary cause of climate change, contributing more than 75% of all greenhouse gas emissions and almost 90% of all carbon dioxide emissions. Greenhouse gas emissions cover the earth, trapping the heat from the sun. These fuels are still largely used in developing countries for energy and transportation. A major challenge in countries in Africa is fossil fuel price hike. More than 98% of transportation in developing countries, especially Africa, is powered by fossil fuels, although there is huge reduction in the usage of fossil fuels in energy sector in developing countries. However, the rural dwellers still use fossil fuels for cooking, lighting, and pocket of other domestic usage.

Electricity and heating account for about 31% global source of greenhouse gas emissions. This is followed by transportation having 15%, manufacturing with 12%, and agriculture and forestry having 11% and 6%, respectively. Therefore, the impact of climate change on energy and transport cannot be overlooked for nations that want speedy industrialization while maintaining net zero emission. Domestic and industrial use of energy hitherto were key drivers of global emissions. Kerosene used in lighting, cooking, and heating is a major contributor of air pollution and attendant health implications. However, deployment and acceptance of clean energy sources such as solar have brought a remarkable reduction in emissions.

Transportation is another driver of air pollution in developing countries. Approximately one-fifth of global CO_2 emissions are attributable to transportation. The main driver of greenhouse gas emissions in the USA is transportation, and CO_2 emissions account for approximately 97% of all transportation-related greenhouse gas emissions as released from the congressional budget office (CBO).

The Director-General of IRENA, Francesco La Camera, proposed that it is time to hasten the switch to a future powered by renewable energy. The likelihood of staying on the 1.5°C path will eventually be eliminated by anything less than radical and immediate action. The energy and climate crises are intertwined, and this has dramatically revealed the flaws and vulnerabilities of an economy that is highly dependent on fossil fuels. Promoting the switch to renewable energy sources is a strategic move that will benefit local people and communities by creating affordable energy, jobs, economic development, and a resilient environment.

If humanity wants to maintain energy security while accelerating the transition to net-zero, Dr. Fatih Birol, Executive Director of the International Energy Agency, believes there is an urgent need to react to the increasing effect of climate change on energy systems. To encourage investment, this calls for long-term planning and audacious policy measures, which must be supported by thorough and trustworthy weather and temperature data. By 2050, solar energy will account for the majority of the world's electricity requirements, which will continue to rise as electrification becomes a strategic tool for achieving net zero objectives. African nations have the opportunity to tap into untapped potential and become dominant participants in the market.

A shift toward clean energy using solar, wind, tidal, hydropower will ensure a positive impact of climate change. Biomass and other clean energy can be used for sustainability. For transport, airbus and other companies have started using hybrid aircraft to reduce greenhouse gas emission. Electric vehicles are reducing the emission produced from road transportations.

There is a growing demand for electric vehicles (EVs). The market for electric cars is one of the most active in the renewable energy sector. In 2021, the number of EV sales increased from the previous year to reach a new high of 6.6 million. Only 120,000 electric vehicles were delivered globally in 2012. More than that many were distributed each week in 2021. In 2021, the market proportion for electric vehicles was nearly 10% worldwide, up from just 2% in 2019. With this, there are now approximately16.5 million electric vehicles on the road worldwide, which is three times more than there were in 2018. With two million electric vehicle sales in the first quarter of 2022, a 75% increase over the corresponding period in 2021, the market for electric vehicles has continued to grow rapidly.

9.2.1.1.2 Impact of Climate Change on Energy and Cooking

The second biggest contributor to climate change after carbon dioxide, cooking over open flames and inefficient wood stoves, produces 25% of the world's black carbon emissions, according to the Clean Cooking Alliance. About six to six and a half of the overall emissions are attributed to cooking.

Crop yields can be reduced, livestock can be killed, and food transportation can be hampered by floods, droughts, stronger storms, heatwaves, and wildfires. Staple cereals such as wheat and rice may become less nutrient-rich as carbon dioxide levels rise as a result of the actions of humans.

Meals are cooked using a range of heat sources, such as gas, wood, and electricity. Cooking with any of these heat sources can pollute the air indoors. Stoves powered by natural gas and propane have the potential to release harmful airborne contaminants like formaldehyde and carbon monoxide that are toxic to both humans and animals. Cooking food releases a variety of gases and particulates into the air, both inside and outside. These emissions are produced when cooking oils and additional ingredients are heated, but they are very dependent on the circumstances and processes used during preparation. Meal with the greatest carbon footprint is beef. It is due to the requirements for farming and raising livestock. The quantity of feed needed for cattle used in the cattle sector must be grown on-site. Moreover, they generate a huge quantity of methane.

The following are some ways of reducing carbon footprint impact in the food industry: avoid using plastic containers, eat more fruit and vegetables instead of beef, or at least reduce your meat consumption. Think about the culinary techniques as well. When possible, buy locally, and eat healthy food.

9.2.1.1.3 Impact of Climate Change on Energy and Agriculture

Crop outputs decline because of droughts and floods. Floods and droughts can damage crops and eradicate the food security while interrupting farming operations and displacing workers as severe weather events become more frequent and more intense. Fuel supply, energy output, and the physical robustness of present and future energy infrastructure are all directly impacted by climate change. Droughts and heat waves are already straining the capacity of current energy production, making it even more crucial to cut back on fossil fuel pollution. Extreme weather conditions like flooding, scorching heat, and drought have degraded the soil, resulting in poor crop yields.

Farmers could adapt their way of life because of declining output from agriculture, particularly in rural areas.

9.2.2 Industrialization and Global South

As a result of the industrial revolution, the global market grew more connected and interdependent. This propensity set up an advantageous environment for what is now referred to as "globalization." In the South, mobilization of resources was more challenging due to a diminished industrial base, fewer rail lines, and an agricultural economy centered on slave labor. The Union had a significant advantage over the Confederacy in terms of industry, railroads, and personnel as the conflict stretched on. Even while New England was mainly excluded from the early American industrial revolution, it eventually expanded to the West and, once the second industrial revolution took place in the late 19th century, to the South, agriculture and handicrafts-based economies were replaced by economies based on large-scale industry, automated manufacturing, and the factory system during the industrial revolution. Current businesses become more profitable and effective because of new equipment, power sources, and organizational techniques.

The 2022 Africa Industrialization Index, issued recently by the African Development Bank, confirms that South Africa, Morocco, Egypt, Tunisia, and Mauritius are the continent's most industrialized economies. The Africa Industrialization Index (AII) research evaluates the development of 52 African nations across 19 important variables on a country-by-country basis. According to a recent report by the African Development Bank, the African Union, and the United Nations Industrial Development Organization (UNIDO), 37 of the 52 African countries have advanced their industrialization over the past 11 years. The 19 indicators of the index include measures of the efficiency of manufacturing, capital, labor, business environment, infrastructure, and macroeconomic stability. The index rates the industrialization of African nations on a trio of criteria: performance, direct causes, and indirect causes. The investments like labor and capital, and how they are used to propel industrial progress, are examples of direct determinants. The enabling environmental factors such as macroeconomic stability, reliable institutions, and infrastructure are indirect determinants. Throughout the 2010–2021 era, South Africa kept a relatively high rating, closely followed by Morocco, which retained second place as of 2022. Egypt, Tunisia, Mauritius, and Eswatini make up the final three places in the top six for the corresponding time period.

9.2.3 Energy Security in the Global South

A sizable portion of the population in the Global South still lacks access to electricity, which is essential for reducing poverty by generating jobs. Energy security, according to the IEA, is the continuous availability of energy sources at competitive prices. There are several facets to energy security; long-term energy security is primarily concerned with making timely investments to meet energy needs as they arise in line with the state of the economy and the environment. Energy security depends on four factors: availability, accessibility, affordability, and dependability. A nation needs to have exposure to a stable and varied mix of energy sources, such as fossil fuels, nuclear power, and renewable energy, to attain energy security.

A variety of tactics can be used by nations to try to attain energy security. These tactics include utilizing resources in one's possession to come as near to complete self-sufficiency as possible, importing energy to supplement one's own supplies from dependable suppliers, obtaining energy from a variety of sources. The main forces behind the global energy shift are climate change and energy security.

9.2.3.1 Impact of Corrosion on Industrialization and Energy Security in Global South

The global economy is entering an unprecedented period marked by technological and environmentally friendly advancement. Multiple threshold effects of the level of advancement in digital industrialization on the amount of time of the energy supply chain exist, and the results are statistically significant; in other words, the growth of digital industrialization positively affects the length of the natural gas supply chain once it reaches or crosses the critical threshold. According to the results of the heterogeneity study, the energy supply chain has been affected differently by digital industrialization depending on the subsectors, the area, and the developmental stage.

In some instances, the surroundings used to produce energy are exceedingly severe; for instance, in the nuclear industry, materials come into touch with high-temperature coolants while being exposed to radiation. Some hydrocarbons, like natural gas, must be extracted from deep, hot, and sour wells in the fossil industry. Corrosion problems seem to be less pervasive or more fundamental to the issue of power generation in other areas, such as the situation of the renewable energy sector. The energy created during nuclear fission is being used to generate power in about 30 nations. Different nations create varying amounts of electricity using nuclear power. Around 75% of the nation's electricity consumption is derived from nuclear power in some nations, such as France. It is over 20% in the USA, while it is about 2% in countries like China [2]. In power plants, corrosion causes expensive repairs, protracted maintenance, material losses, subpar performance, and, if unchecked, failure.

9.3 CONCEPTUAL AND THEORETICAL PERSPECTIVES OF INDUSTRIALIZATION, CLIMATE CHANGE, AND ENERGY SECURITY

Industrialization is a shift from a consumer mindset to producer by providing conducive environment for business and industries to grow. Energy security is key to achieving uniform industrialization in a region or country. Energy security is the continuous and uninterrupted availability of energy to a specific region or country. Energy security emphasizes availability, affordability, and sustainability. Energy contributes to greenhouse gas emissions that deplete the ozone layer. The continuous depletion of the ozone layer causes a shift in global climate pattern known as climate change. Since the preindustrial era (1850–1900) an increment of 1.1°C has been experienced in global average surface temperature. This has resulted in frequency and severity of climate shocks globally, as well as in heat waves, flooding, severe storms, and wildfires, among other weather extremities. IPCC report predicts further increment of 4°C of global warming by 2100 if greenhouse gas emissions are not curtailed. Although 189 countries have committed to reduce CO_2 emissions by 30% in 15 years until 2030, global CO_2 emissions continued to increase since the 2015 Climate Accord by 2.3% to 36.3 billion metric tons in 2021—the highest level in history.

A report by International Monetary Fund (IMF) outlines the current global perspective for energy security and climate change as it relates to industrialization. Empirical analysis by IMF of 39 countries in Europe over the period 1980–2019 shows that climate change and energy security are issues having two faces of the same coin (IMF, 2020). IMF empirical analysis shows that increasing the share of nuclear, renewables, and other non-hydrocarbon energy and improving energy efficiency could lead to a significant reduction in carbon emissions and improve energy security. An affordable, available and sustainable, eco-friendly energy source will drive industrialization, as energy security is the bane of industrialization.

Numerous efforts and policies have been made in Global South to improve the energy supply, reduce greenhouse gases, and encourage industries and business to grow. However, lack of awareness and the will to implement them have hindered things.

9.4 PERSPECTIVES AND TRENDS OF INDUSTRIALIZATION, CLIMATE CHANGE, AND ENERGY SECURITY IN GLOBAL SOUTH

Consequent to human activities, greenhouse gas emissions into the atmosphere have dramatically increased. Burning fossil fuels, such as coal and oil, has increased the quantity of greenhouse gases in our atmosphere since the industrial revolution began around 1750. Global energy security is jeopardized because of climate change [3]. Fuel availability, energy production, and the physical robustness of present and future energy infrastructure are all directly affected by climate change.

The trends of global industrialization are moving toward the 5th industrial revolution away from the 4th industrial revolution (4IR). Many countries in the Global North have achieved growth using the 4IR. China, USA, and Germany continue to dominate in attainment of several of the technologies that make up the 4IR. Yet, most countries in the Global South continue to battle with 2nd and 3rd industrial revolutions. Countries such as Nigeria and Benin Republic still suffer from epileptic power supply, 3G or 4G internet data access, whereas wireless technology is way ahead of 5G. The success of 4IR is based on stable power supply, high-speed internet, and wireless technology. Yet the solution continues to elude most of the countries in the global south.

Energy security in the Global South is still a distant goal. Cooking with firewood, lighting of kerosene lantern, and driving of fuel-powered cars that were manufactured over 20 years ago are till prevalent in the Global South, whereas countries in the Global North have advanced to small modular reactors, clean coal, and hydrogen, among other technologies. The deployment of emerging technology such as 3D printing to reduce the cost, improve the efficiency, and optimize key variables for energy generation, distribution, and transmission has gained ground in the Global North yet limited is known and utilized in most countries of the Global South.

9.5 ISSUES AND CHALLENGES OF INDUSTRIALIZATION, CLIMATE CHANGE, AND ENERGY SECURITY IN GLOBAL SOUTH

There is a need for assessment and policies to drive industrialization, energy security, and climate change: research in Global South is often not driven by the need. This has resulted in waste and duplication of resources.

Political will: Policies are available and common in Global South. However, the political will to implement them is lacking. Most are well crafted and drafted but are lying on the shelves of various ministries without full implementations.

The following are some of the ways that will improve the lot of Global South in industrialization, energy security, and climate change: development of materials for industrialization, encouraging technology that reduces greenhouse gas emission in Global South, aggressive research and development of materials for energy security, especially polymer, nanostructured materials, and organic energy materials.

9.6 FUTURE OUTLOOK OF INDUSTRIALIZATION, CLIMATE CHANGE, AND ENERGY SECURITY IN GLOBAL SOUTH

Policy makers, researchers, and stakeholders should develop adequate strategy for research and development and policy implementations that will ensure only energy sources that reduce greenhouse gas emission are used in Global South to curb climate change. Such strategy should ensure products and services developed are affordable and available. Researchers and innovators should be encouraged through funding, capacity building, and linkage to international community for aggressive research in clean energy. We need to establish laboratories that perform research and development on solar, wind, hydro turbine, and nuclear in Global South and use of emerging technologies such as 3D printing and artificial intelligence to reduce the cost and improve efficiency of clean energy. Moreover, deployment of advanced manufacturing technology in the production and research system of Global South will be fruitful.

9.7 SUMMARY

Corrosion has a major role to play in the industrialization of a region or country. A lot of time and resources are spent in combating the menace of corrosion, thereby resulting in slow pace of industrialization. Energy security in the Global South is threatened by corrosion and dilapidated infrastructures. This has resulted in alternate sources such as clean and renewable energy. However, they are still affected by corrosion. This chapter discussed the link between corrosion, industrialization, energy security, and climate change in the Global South. Global South is a collective name used for regions of the world that are characterized by large countries with developing economy. It was discovered that the attainment of developed economy by countries in the Global South will be possible with deployment of appropriate technology, policy that combats greenhouse gas emission, and corrosion which will result in speedy industrialization and energy security of the regions. Adequate investment in research and development will accelerate even development and industrialization of the Global South.

REFERENCES

1. Popovich, N., & Plumer, B., 2021. Who has the most historical responsibility for climate change? Retrieved from www.nytimes.com/interactive/2021/11/12/climate/cop26-emissions-compensation.html
2. Rebak, R.B., 2013. Introduction to corrosion in energy production. *JOM*, *65*(8), pp.1021–1023.
3. Cevik, S., 2022. *Climate Change and Energy Security: The Dilemma or Opportunity of the Century?* An International Monetary Fund report. Retrieved from www.imf.org/-/media/Files/Publications/WP/2022/English/wpiea2022174-print-pdf.ashx

10 Corrosion and WASH/ Smart Coating

10.1 INTRODUCTION TO WATER SANITATION AND HYGIENE (WASH)

Sanitation, cleanliness, and access to clean water are essential for human health and welfare. Safe WASH is important for livelihoods, education, and dignity in addition to being a requirement for good health. It also helps build strong communities with wholesome surroundings. Sanitation and access to clean water are fundamental human rights. However, billions of people worldwide do not now have access to clean water, and hundreds of millions do not have access to the sanitary facilities they require. Untreated excrement pollutes groundwater and surface waterways used for drinking water, irrigation, showering, and domestic functions, which can lead to ailments including diarrhea and harm to health.

A total of 844 million individuals globally lack access to clean drinking water and 2.3 billion lack access to latrines or other forms of basic sanitation. Among the top reasons for death for children under the age of five are polluted water and inadequate sanitation. The ability to access WASH services is necessary to improve the standard of living in the context of sanitation, advance public health, apply the human right to water and sanitation, ease the load of women's water collection, lower the risk of abuse against women, enhance educational and medical outcomes at institutions of higher learning, and lessen water pollution. A crucial element of water security is access to WASH services. Access to clean water and sanitation for everybody is a crucial issue in worldwide development.

The World Health Organization (WHO) estimates that poor sanitation causes 700,000 annual child deaths from diarrhea, mostly in poorer nations. According to the Bill and Melinda Gates Foundation, children who experience persistent diarrhea may suffer long-term harm to their mental and physical growth. However, due to a number of complication variables, it is difficult to gather accurate scientific data concerning health outcomes that result from better access to WASH. To more accurately examine WASH health outcomes, academics contend that longer-term technological efficacy studies, fuller analysis of sanitation programs, and research of the cumulative effects of various interventions are necessary.

Urban slum service delivery, WASH breakdowns (such as leaking water distribution systems), water contamination, and the effects of climate change are all

DOI: 10.1201/9781003441151-10

difficulties. Integrated water resources management (IWRM), national WASH plans and monitoring (including gender mainstreaming), and, lately, enhancing the climate resilience of WASH services, are planning strategies for improved, more dependable, and equal access to WASH. Water management systems with an adaptive ability can better withstand the adverse effects of climate-related events and boost climate resilience. Small urban utilities to national governments are just a few of the stakeholders who need access to trustworthy information on the local climate and any anticipated changes brought on by global climate change.

10.1.1 Water, Treatment, and Impact on Humanity

Humans need to drink an appropriate quantity of water daily to stay alive. However, this differs depending on a person's age, gender, and place of residence. An adult male typically requires 3 L (3.2 quarts) of fluid per day, whereas an adult female typically requires 2.2 L (2.3 quarts) per day. The body can survive without food for about a week but may result in death if there is no water for days. Water makes up about 50–75% of the body. The ten water poorest countries in the world for access to potable water are all located in Africa. In the Democratic Republic of the Congo and Niger, respectively, 54% of the population lacks access to clean water. About 50% of Ethiopians lack access to basic water utilities. In Somalia, 44% of the population lacks access to basic water services. 43% of Angolans lack access to basic water services. Only 3% of the water on Earth is freshwater; the majority is salty. The United Nations (UN) estimates that over 66.67% of people experience water scarcity for a minimum of one month each year. Water shortage is getting worse, especially in low-income nations, due to a growing world population, poor infrastructure, and climate change. When there is inadequate water to satisfy all needs, including those of ecosystems, there is a physical water shortage. Arid regions, including Central and West Asia and North Africa, frequently have a physical water shortage.

10.1.1.1 Water Treatment

This is the process of making water clean and safe for drinking or use by livestock and humans. The five most common processes of water treatment include ion exchange, filtration, disinfection method, reverse osmosis, and activated carbon filters as shown in Figure 10.1.

They are used in whole or combined depending on the level of treatment.

10.1.1.1.1 Reverse Osmosis

Reverse osmosis (RO) produces clean, excellent-tasting water and is regarded by many as one of the most efficient water filtration techniques.

When pressure pushes water across a semipermeable membrane, RO eliminates pollutants from untreated water, also referred to as feed water. To produce clean drinking water, it moves from the highly concentrated side of the RO membrane – which has more impurities – to the least concentrated side which has fewer impurities. The permeate is the name for the produced fresh water. The waste or brine is

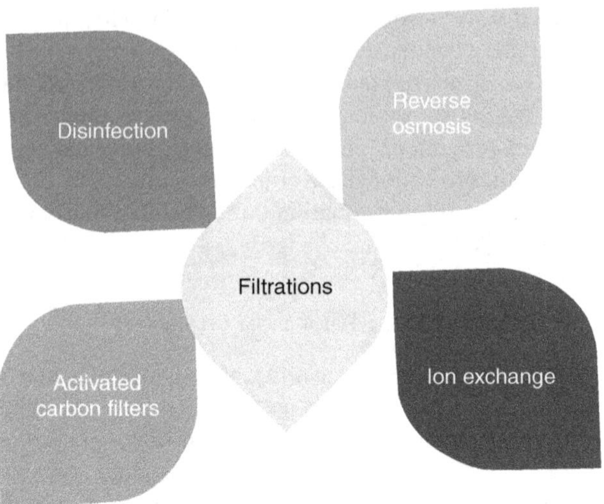

FIGURE 10.1 Classifications of water treatment types.

leftover concentrated water. Small pores in a semipermeable membrane facilitate the passage of water molecules but impede the passage of pollutants.

Osmosis involves the concentration of water as it moves across the membrane to achieve equilibrium on each side. Contaminants cannot get through the membrane's less concentrated side thanks to reverse osmosis. In reverse osmosis, the salt is left behind leaving clean water when pressure is introduced to a volume of saltwater. Freshwater will pass through the semipermeable barrier and dilute the concentrated solution during natural osmosis. Reverse osmosis is a process in which pressure is introduced to a concentrated aqueous solution to drive the water molecules through the membrane and into freshwater.

Numerous uses for RO systems exist, such as faucets, aquariums, whole-house, and restaurant filtration. Whatever the initial quality of your water, there is probably an RO system that will meet your demands. A sediment filter, pre-carbon block, reverse osmosis membrane, and post-carbon filter are commonly used in reverse osmosis. To keep the preceding filters from being clogged, the sediment filter eliminates the biggest particles, such as dirt, sand, and rust. The process of reverse osmosis includes prefiltration which occurs before the water enters an RO system. After that, dissolved particles – even those that are too minute to be detected with an electron microscope – pass through the reverse osmosis membrane and are eliminated from the water. Water moves to the reservoir after filtering, where it is kept until required. Figure 10.2 presents the working process of a reverse osmosis water treatment.

An RO system's main component is the RO membrane. However, an RO system also has additional filters. There are three, four, or five filtration phases in an RO system. Besides the RO membrane, every reverse osmosis water system also includes a sediment filter and a carbon filter. Depending on whether they are used before or after the membrane, filters are referred to as pre-filters or post-filters. There are one

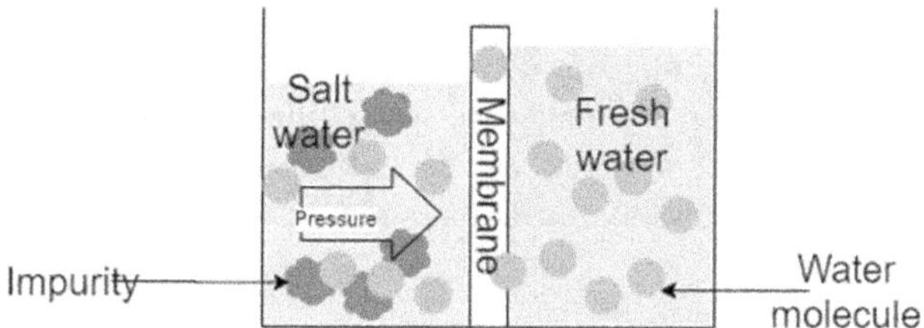

FIGURE 10.2 A schematic of the working operation of reverse osmosis water treatment.

or more of the following filters in each type of system viz sediment filter that lowers the number of debris, dust, and rust. Volatile organic compounds (VOCs), chlorine, and other impurities that give water a foul taste or odor are reduced using carbon filters. Up to 98% of total dissolved solids (TDS) are removed by semipermeable membranes.

Reverse osmosis is durable with a lifespan of about 15 years. The filters (pre-filters and post-filters) are replaced between six months and one year. However, the membrane should be replaced often every three to four years.

The type of reverse osmosis to be installed, the pretreatment, installation cost, maintenance cost, and need for extra equipment all affect the cost of reverse osmosis. Similarly, certain things need to be considered before setting up reverse osmosis. They include water source, water pressure and water demand, reverse osmosis style, budget, installation, remineralization, and proposed efficiency. Any RO system installation is an investment that comes with ongoing maintenance requirements, such as buying new filters and membranes. Whether you need an RO system under the sink or for an entire home, as well as how many additional components you require, will affect the overall cost. Whole-house RO systems are significantly more costly and complicated. Depending on the design, under-sink RO systems typically cost $200 to $400. If the system has a booster pump, permeate pump, storage tank, or pretreatment, the price goes up. If the well water has been cleansed for bacteria and other debris that may block the membrane, RO systems can handle both city and well water. Once the water exits the storage tank and gets dirty with heavy metals like arsenic, an ultraviolet (UV) system may be required to purify it. Nutritionists recommend that each person drink half their body weight in ounces each day. A 140-pound person has to consume 70 ounces of water each day. Moreover, the amount of water consumed by refrigerators, ice makers, and all other appliances should be considered. The volume of water generated daily by each reverse osmosis system is listed in GPDs (gallons per day). Water pressure is used in RO to push clean water through the membrane. A RO booster pump is needed to raise the pressure for RO if the water pressure in a home is less than 40 psi. A tank's water pressure needs to be set high enough for an RO system to function for water obtained from well water. The amount of pressure required for the RO system to create clean water is also increased when there is a high concentration of dissolved solids in the water. Three main cartridge types are available for reverse osmosis systems: quick-change, drop-in, and standard.

RO removes dissolved particles from water while lowering pH or reducing alkalinity. The water is slightly acidic due to the decreased pH. Adding an acid-neutralizing filter, such as a calcite filter or an alkaline water filter, raises the pH and imparts a mineral flavor to acid water, which is corrosive by nature. Through many phases of remineralization, the filter dissolves calcium and magnesium in water and minimizes scale, bacteria, and silt.

10.1.1.1.2 Disinfection Method

Water treatment facilities may add one or more chemical disinfectants (such as chlorine, chloramine, or chlorine dioxide) after filtration to remove any lingering parasites, bacteria, or viruses. Water treatment facilities make sure the water has low concentrations of the chemical disinfectant before it leaves the facility to help keep water safe as it goes to homes and businesses. The extra disinfectant eliminates bacteria that are present in the pipes that run from the water treatment facility to the tap. UV radiation can be used to treat water in water treatment plants alongside chlorine, chloramine, or chlorine dioxide. In the treatment plant, ozone and UV light sterilize water effectively, but they do not keep bacteria from multiplying.

10.1.1.1.3 Ion Exchange Water Treatment

In the process of treating water, ion exchange involves replacing unwanted ionic pollutants with another ionic material. Water softening, deionization, demineralization, and de-alkalization are the four ion exchange techniques. To make water safe, pollutants must be eliminated.

In a huge aeration tank, untreated water is initially gathered and aerated by pumping compressed air through perforated pipes. Aeration eliminates CO_2 and foul odors. Additionally, it eliminates metals like iron and manganese by causing their corresponding hydroxides to precipitate. A settling tank is used to keep the aerated water for 10–14 days. Within 24 hours of storage, 90% of the suspended particles settle down, making the water clear. During storage, some heavier dangerous compounds also settle down. Akin to how harmful bacteria progressively die, the number of bacteria in the first 5–7 days of storage drops by 90%. Microorganisms oxidize organic materials in water that are present during storage. Similarly, during storage, NH_3 is oxidized into nitrate by microorganisms. After being fed to the coagulation tank, the water from the storage tank is combined with alum, lime, and other precipitating agents.

When dissolved in water, these precipitating agents produce a precipitate of $Al(OH)_3$. Precipitation accumulates suspended materials from its surface, gradually gaining weight until it settles down. Very light suspended particles that do not settle naturally during storage are removed using this method. Additionally, negatively charged colloidal contaminants get eliminated by Al^{3+} ions and settle if they are present.

The remainder of the pollutants and bacteria are removed using a sand gravity filter, which is used with partially cleared water. A rectangular tank called a sand filter houses a three-layer filter bed with a one-meter-thick fine layer on top. The middle layer has a layer of coarse sand that is 0.3–0.5 m thick. Gravel is present in the bottom layer and is 0.3–0.5 m thick.

The filtered water is collected in a tank at the base of the filter bed. During filtering, a slimy layer known as the crucial layer quickly covers the filter bed. Algae, diatoms, and bacteria that resemble threads make up the vital layer. Microorganisms in the crucial layer oxidize organic and other water-based materials during filtration. For instance, NH_3 is converted to nitrate when it is present. Filtration of microbial cells is assisted by the vital layer as well.

If the water has an offensive odor, activated carbon can be added to the filter bed to remove it.

Disinfectants are then used to purify the filtered water. Examples of disinfectants include fluoride and chlorination. Disinfectants eliminate both pathogenic and non-pathogenic microorganisms from water. Water flows from the overhead tank to the home for domestic usage.

10.1.1.1.4 Filters for Water Treatment

Filters trap various sizes of particles and ions from untreated water to make it safe and clean for drinking. Filters eliminate germs and dissolved contaminants such as dust, chemicals, parasites, bacteria, and viruses. There are about eight types of filters namely alkaline and water ionizers, activated alumina, ceramics filters, distillation, carbon block activated carbon, sand and sediment mesh, UV light, and reverse osmosis as depicted in Figure 10.3.

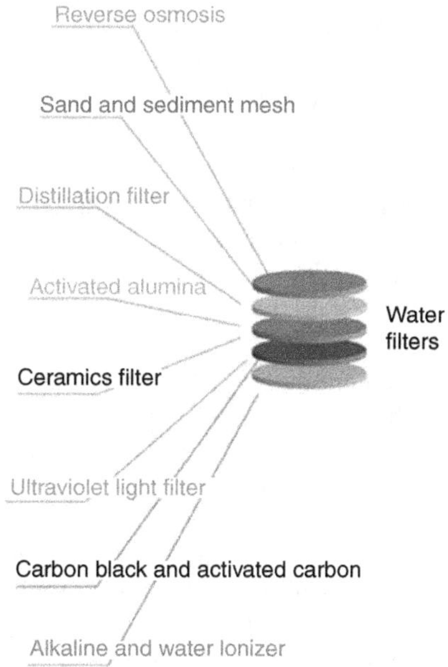

FIGURE 10.3 Classification of filters for water treatment.

Activated Alumina

The finest filter to use to eliminate fluoride or arsenic from water is activated alumina (AA). This highly porous ceramic material is made of aluminum oxide and has a high capacity for adsorption – not absorption. Chemicals that are drawn to AA's activated porous sphere are not just kept inside the pores; they also form bonds with one another. Because of this property, AA is an economical solution for removing metals and poisons from polluted water without running the danger of these materials getting into the environment. More dangerous substances can be removed from water as long as AA is in contact with it. Toxic waste remediation, industrial uses, water treatment facilities, and residential water filtration can all benefit from the use of activated alumina.

Ceramic Filters

To remove dirt, debris, and germs from water, ceramic water filters (CWFs) use the microscopic pores of the ceramic material, which is a cheap and efficient type of water filter. Consequently, portable ceramic filters are excellent for usage in underdeveloped nations and are frequently used in backpacking. Any material larger than the microscopic pores on a ceramic surface will be trapped by the pores, keeping it out of your water. These filters are cheap, easy to use, simple to install, and do not require energy. CWFs can also eliminate bacteria and stop mold and algal growth if they have been treated with silver. They are ineffective at removing viruses, though, because they cannot move any quicker than they can filter.

Similar to other water treatment methods, ceramics filter eliminates particles that are wider than the holes in the filter material. Bacteria, protozoa, and microbial cysts are often eliminated. However, as viruses can fit through filters and reach the "clean" side, they are typically ineffective against them. Silver may be added to CWFs in a non-leaching form. The silver aids in destroying or immobilizing germs and hinders the growth of mold and algae in the filter's body.

Chemical pollutants are not specifically removed by ceramic filtration. To decrease organic and metallic impurities, some manufacturers (particularly ceramic candle filters) include a powerful activated carbon core inside the ceramic filter cartridge.

Chemicals like chlorine are absorbed by the active carbon. Active carbon filters must occasionally be updated as the carbon becomes blocked with contaminants. Candle-type and pot-type CWFs are the two most popular varieties. In ceramic filter systems, a porous ceramic filter is either attached to or placed over a ceramic or plastic container. A top container is filled with contaminated water. It enters the container below after passing through the filter(s). The lower outlet frequently has a tap installed. The top half of the unit will hold any impurities that are wider than the tiny holes in the ceramic construction. A gentle brush and clean water, hot water, and soap can be used to scrub the filter(s) clean.

Compared to inserts and pots, ceramic candles have mechanical, operational, and manufacturing advantages. Filter candles make it possible to utilize robust metal and plastic containers, which lowers the risk of hygienic failure. Additionally, there are in-line ceramic filters that filter drinking water that enters through home plumbing as well as portable ceramic filters that use manual pumping. Similar to the clay pot filter, these filters can be cleaned by reverse flow, which forces clean water through the filter backward to remove any impurities from the ceramic pores.

Hairline cracks and cross-contamination pose the biggest threats to the effectiveness of all types of ceramic filtration. The fragile nature of ceramic materials can cause fine, hardly perceptible cracks if the device is dropped or improperly mistreated, allowing bigger particles to get past the filter. To increase the brittleness and fracture toughness of these clay CWF materials, research is ongoing to alter the clay/sawdust ratios during manufacturing. The filtering will not work if the "clean" water side of the ceramic membrane comes into touch with unclean water, hands, cleaning cloths, etc. The clean side of the filter should be completely cleaned before reuse if such contact occurs.

UV Light

One of the most secure and efficient methods for eliminating viruses, germs, protozoa, and parasites from water is UV light. High-frequency UV light radiation strikes water and travels through a glass component. The cost of this filtration process is on the higher end and it takes electricity to operate. The optimum time to utilize UV lamps as water purifiers is at the end of the water treatment process when the water has been most thoroughly filtered. This is because non-organic impurities cannot be removed by UV light, which reduces the effectiveness of UV lamps.

Carbon Black and Activated Carbon

These are frequently used in home water filters, such as those on faucets, under the sink, or in a water pitcher. They are also referred to as carbon filters or activated charcoal. Pollutants are effectively drawn to and adsorbed by activated carbon filters, which remove them from the water. This product is cost-effective and does not require energy to operate; however, it is ineffective at removing minerals and dissolved organic debris.

Alkaline and Water Ionizers

One of the healthiest water purification methods is alkaline or ionized water, which is thought to provide advantages like better skin, better bones, and digestive health, a lower risk of cancer and heart disease, greater hydration, and better taste.

Ionized water's electrically charged minerals separate the alkaline from the acidic water, making the water softer and better for drinking. Alkaline ionizers do not have good filtration capabilities; thus, they should be used in conjunction with other water purifiers.

Distillation

Water is heated into steam during the distillation process, which is subsequently cooled to bring the water back to a liquid state in a sterile container. The procedure eliminates microorganisms while also enhancing the flavor and odor of treated water. The distillation procedure can be completed using a variety of countertop tools to obtain safe drinking water. The procedure is extremely sluggish and does not win for efficiency because it needs energy or a heat source to function.

10.1.2 Sanitation

According to the Centers for Disease Control (CDC), basic sanitation is the capacity to sustain hygienic conditions through services like garbage pickup, management of industrial and hazardous waste, and wastewater treatment and disposal. It also refers to having the necessary facilities for the safe disposal of human waste (feces and urine).

Safe drinking water, the purification and evacuation of human waste, and other circumstances linked to public health are referred to as sanitation. Sanitation includes washing hands with soap after using the restroom and avoiding contact with human waste. Sanitation systems work to safeguard public health by creating a hygienic setting that will halt the spread of disease, particularly through the fecal–oral route. For instance, diarrhea, a major contributor to child malnutrition and stunted growth, can be decreased with proper sanitation. Ascariasis, a kind of intestinal worm infection or helminthiasis, cholera, hepatitis, polio, schistosomiasis, and trachoma, to mention a few, are just a few of the diseases that are easily spread in communities with poor sanitation.

There are numerous sanitation technologies and methods available. They include ecological sanitation, emergency sanitation, environmental sanitation, onsite sanitation, and sustainable sanitation. Other instances include community-led complete sanitation and container-based sanitation. Human excreta and wastewater are captured, stored, transported, treated, and either reused or disposed of as part of a sanitation system. Activities for recycling wastewater and excreta may concentrate on recovering nutrients, water, energy, or organic materials. "Sanitation workers" are those who are in charge of operating, maintaining, or emptying a sanitation technology at any point in the sanitation chain.

Sanitation is absent when there is a lack of it. It refers to a lack of restrooms or restrooms that are not clean enough for someone to voluntarily use them. Open defecation (and open urination, though this is of less significance) are frequently the result of poor sanitation, which has major public health implications. In addition to 660 million people lacking access to safe drinking water as of 2015, it is estimated that 2.4 billion people still lacked appropriate sanitary facilities. For monitoring, improved sanitation is a term used to group different types of sanitation. It speaks to how human waste is handled at the domestic level. It was created in 2002 to track progress by the Joint Monitoring Program (JMP) for Water Supply and Sanitation of United Nations International Children's Emergency Fund (UNICEF) and WHO.

An effective sanitation system that is long-lasting and that meets specific standards is known as sustainable sanitation. The complete "sanitation value chain" is taken into account when designing sustainable sanitation systems, including user experience, wastewater and excreta collection technologies, waste transportation or conveyance, waste treatment, and reuse or disposal. The Sustainable Sanitation Alliance (SuSanA) defines "sustainable sanitation" as having five characteristics. Systems must be technically sound, institutionally appropriate, and economically and socially acceptable while also protecting the environment and natural resources.

10.2 CORROSION AND WASH

Corrosion affects every area of WASH. Most of WASH facilities face corrosion and aging. Most of the taps used for toilets and sanitation are aging. These sanitation-related components have similar problems. The first is that a lot of them are made of concrete or brick and mortar, both of which are easily corroded by chemicals, water, and the environment. Additionally, the steel rebar used to stabilize the large slabs of concrete can become displayed through a crack and start to rust. As the rust grows, it causes the concrete to crack even more. Sanitation systems' tendency to be damp is a second prevalent issue. This results from both the humid environment and the liquid components passing through or being held. Many concrete coatings are not built to last.

In Singapore, 900,000 government-constructed and managed housing units are home to 85% of the population. Many of the 1960s-era concrete garbage chutes are currently deteriorating. These hoppers allow the garbage from the residences to fall into collecting containers, which are then taken away. For convenience, this type of garbage disposal is installed everywhere, but when corrosion eats away at the edges, it may start to create a bad environment for the growth of bacteria, unpleasant aromas, and vermin.

10.3 STATE OF AFFAIR OF WASH

The statistics are shocking with almost half of the world's population –3.6 billion people –using sanitation systems that do not process human waste, and 2 billion people lack access to clean drinking water. Millions of children and families lack access to sufficient WASH services, such as soap for hand washing. The results are frequently fatal. At least 1.4 million people every year, the majority of these kids, pass away from preventable diseases associated with contaminated water and subpar sanitation. For instance, cholera is now prevalent in nations where there have not been epidemics in decades.

The UN Sustainable Development Goal (SDG) 6 –providing access to water and sanitation for all by 2030 still needs to be accelerated in several nations, according to the Global Analysis and Assessment of Sanitation and Drinking Water (GLAAS) report from WHO and UN-Water. This is because most nations are still behind in achieving the goal. Only 25% of nations are on track to meet their national sanitation targets, while 45% of nations are on track to meet their nationwide defined drinking-water coverage targets. Less than one-third of nations reported having enough human resources to complete essential drinking water, sanitation, and hygiene (WASH) tasks. The GLAAS 2022 report summarizes the most recent WASH system status in over 120 countries, and represents the largest data collection ever undertaken from the broadest range of nations. Despite an increase in WASH budgets in certain nations, more than 75% of nations reported having inadequate funds to carry out their WASH plans and objectives.

Most WASH policies and plans fail to take climate change threats to WASH services or the climatic resilience of WASH technologies and management systems into account, according to the GLAAS data. Only slightly more than two-thirds of nations have WASH policies formulated that target populations that are most affected by climate change. Only about one-third, though, explicitly fund these populations or track their progress.

In half of all healthcare institutions, where good hygiene habits are most important, there is no access to water, soap, or alcohol-based hand sanitizer. Everyone has a right to clean water, good sanitation, and good hygiene, but many people lack these things. For everyone to have access to securely managed WASH services by 2030, global progress must at least treble from its current rates. To safeguard people's health and prospects, progress must be made even more quickly in unstable environments and the world's poorest nations.

10.3.1 The Role of Government in Driving WASH

The government needs to provide enabling leadership by; creating a strategy for boosting political commitment to responsibly managed drinking water, sanitation, and hygiene, which includes connecting with civil society organizations and reaching out to leaders at all levels of government. Creating a plan for enhancing the institutions and governance needed to provide these services, such as by setting up independent regulatory organizations that uphold health-based standards and frequently disseminate findings.

Create specific policy goals to direct funding and financing choices for WASH. Create costed funding and financing plans that consider the requirements of various geographic areas and population groups. To acknowledge the importance of WASH as a public good, increase public investment in it. Encourage service providers to perform better to delight customers and recoup expenses, for instance by lowering service disruptions and water losses and enhancing tariff designs and collection efficiency.

10.4 BRIEF ON SMART COATINGS

Smart coatings are specialized films with predetermined characteristics that enable them to sense and react to external stimuli such as the environment. The coatings' capacity for self-healing and self-repair makes them excellent for applications including surface enhancement, material protection, and corrosion prevention.

Smart coatings blend functionality and aesthetics to provide both common coating functions, such as protection and ornamentation, as well as special functionalities that depend on environmental cues.

Smart coatings are coatings created to provide the surfaces they are applied to functionality. Smart coatings can actively recognize and react to environmental stimuli including pressure, temperature, light, and heat. They can then respond appropriately. Recently, nanomaterials have become an effective method for delivering improved functionality in the creation of such coatings. Smart coatings are available for self-cleaning, antibacterial, self-healing, corrosion resistance, and other purposes.

Due to several factors including rising demand from end-user industries, rising demand for low-maintenance products, and rising desire for superior qualities, the smart coatings market is anticipated to develop till 2027. The high cost of smart coatings and rigorous environmental laws might impede progress.

An object can be given a coating to add color and gloss. Therefore, coatings have two functions: to protect an object and/or to embellish it. By cutting down on inspection hours, maintenance expenses, and equipment downtime, smart coatings can increase a system's efficiency. In addition to prolonging the lifespan of parts and structures composed of corrosive materials, the smart coating also reduces the requirement for repairing corroded regions.

The use of barrier coatings, such as paint, plastic, or powder, is among the simplest and least expensive techniques to prevent corrosion. Epoxy, nylon, and urethane powders stick to the metal surface to form a thin layer. Metal surfaces are frequently treated with plastic and waxes. Coatings offer defense against damage brought on by applications that are prone to wear. They were made to be wear-resistant, decrease friction, guard against caustic/acidic substances and cleaning agents, and boost line efficiency. There are two types of coatings: organic and inorganic.

10.5 SUMMARY

WASH is a term used collectively to describe water sanitation and hygiene. Sanitation, cleanliness, and access to clean water are essential for human health and welfare. Nutritionists recommend that each person drink half their body weight in ounces each day. A 140-pound person has to consume 70 ounces of water each day. The United Nations (UN) estimates that over 66.667% of people experience water scarcity for a minimum of one month each year. The five most common processes of water treatment include ion exchange, filtration, disinfection method, reverse osmosis, and activated carbon filters. Sanitation is the capacity to sustain hygienic conditions through services like garbage pickup, management of industrial and hazardous waste, and wastewater treatment and disposal. Numerous sanitation technologies and methods are available which include ecological sanitation, emergency sanitation, environmental sanitation, onsite sanitation, and sustainable sanitation. About 660 million people lacked access to safe drinking water as of 2015. It is estimated that 2.4 billion people still lacked appropriate sanitary facilities. About 3.6 billion people use sanitation systems that do not process human waste, and 2 billion people lack access to clean drinking water. More than 75% of nations reported having inadequate funds to execute their WASH plans and objectives. Smart coatings are able to actively recognize and react to environmental stimuli including pressure, temperature, light, and heat. From the foregoing, WASH is crucial for the survival of humanity. There needs to be a consented global effort to bridge the huge gap in ensuring all achieve the target of SDG goal 6.

Index

Note: Page numbers in *italic* refer to figures, page numbers in **bold** refer to tables.

D

E

F

G

H

I

R

S

T

For Product Safety Concerns and Information please contact our EU representative GPSR@taylorandfrancis.com
Taylor & Francis Verlag GmbH, Kaufingerstraße 24, 80331 München, Germany

www.ingramcontent.com/pod-product-compliance
Lightning Source LLC
LaVergne TN
LVHW010601110826
845149LV00003B/729

* 9 7 8 1 0 3 2 5 7 8 1 8 7 *